T0174239

Discover Excellence

An Overview of the Shingo Model and Its Guiding Principles

The Shingo Model Series

Shingo Institute
Jon M. Huntsman School of Business
Utah State University
3521 Old Main Hill
Logan, UT 84322-3521 USA

Discover Excellence

An Overview of the Shingo Model and Its Guiding Principles

Edited by
Gerhard Plenert
Former Director of Executive Education
Shingo Institute

CRC Press is an imprint of the
Taylor & Francis Group, an **informa** business
A PRODUCTIVITY PRESS BOOK

CRC Press
Taylor & Francis Group
6000 Broken Sound Parkway NW, Suite 300
Boca Raton, FL 33487-2742

© 2018 by Utah State University
CRC Press is an imprint of Taylor & Francis Group, an Informa business

No claim to original U.S. Government works

International Standard Book Number-13: 978-1-138-62616-4 (Hardback)

Visit the Taylor & Francis Web site at
http://www.taylorandfrancis.com

and the CRC Press Web site at
http://www.crcpress.com

To the love of my life—Renee Sangray Plenert

Where I discovered excellence!!!

And to the Shingo Institute—which strives for a world of excellence

Gerhard Plenert

Contents

SECTION III Knowing Why

SECTION IV   Wrap Up

Preface

In a meeting with a CEO (chief executive officer), Dr. Gerhard Plenert was told, "We have had dozens of consulting companies through here, each telling us they know the best way to solve our enterprise's performance problems. But every one of them only solves a small part of the problem. And what's even more frustrating is that as soon as the consultant leaves, the employees quickly revert back to their previous business practices. It's just been a big waste of money. Are you going to come in here and give me another solution which fixes a part of our problem but leaves us missing the mark? Are you going to give us another quick-fix wonder which only offers a temporary solution?"

Dr. Plenert was quick to jump at the opportunity and stated, "What you're telling me is that all you're getting is a bunch of tools. You're not attacking your problems at the correct level of your organization. You're not building a sustainable solution to your performance problems."

"What do you mean?" asked the CEO.

"The good news is that you've already identified the problem," Dr. Plenert continued. "Your problem isn't that the tools you've been given aren't good tools. It's not that the tools aren't working. The problem is that the tools by themselves don't offer sustainability. So the real question we should ask is, 'Why don't these tools maintain the performance improvements that they demonstrate on a temporary basis?' A second problem is that each solution temporarily fixes part of your performance problems. But none of them fix all the problems."

"So what you're saying is that I need some kind of integrated solution," inserted the CEO. "I understand that. But how do I make it sustainable?"

"The only way to get sustainability is through an enterprise-wide cultural shift. You currently have a culture that has been established through many years of tradition. It's ingrained into the company. They have a 'this is how it's always been done' attitude and they believe that tradition is the best way to do things. They also have a 'we're different than anyone else' attitude which makes them resistant to change. They don't believe anyone else truly understands what they do and how they do it. They're willing to play along with the latest fad that comes down from the executive office, which is how they regard the change initiatives that you've introduced, but as soon as the change agent leaves the building, they revert back to tradition."

"You've captured my interest and curiosity," responded the CEO. "What do you recommend?"

"Let me introduce you to the Shingo Institute (www.shingo.org). It's an organization affiliated with Utah State University. They host an international award in enterprise excellence and they have a training program associated with that award. This training will teach you how to shift the behaviors within your organization so you can successfully transform your enterprise-wide culture. The Institute teaches you how systems drive behaviors, and how only through a shift in these behaviors will you ever be able to achieve a culture that sustains a higher-level of performance."

With that Dr. Plenert and the CEO proceeded to discuss the details of the Shingo methodology (which is the subject of this book), and a series of Discover Excellence workshops were scheduled for the leadership and management of the organization. This was the first step toward achieving enterprise excellence, and that is what this book will teach the reader how to do.

But first, just for fun, one should look at how good an organization is at solving problems. Below is a test created by Einstein. It's called the Albert Einstein Riddle. Albert Einstein wrote this riddle and claims that if a person can solve this "pure logic" problem then that person must be in the top 2% of the intelligent people in the world. He starts with these rules:

1. On a street, there are five houses painted five different colors.
2. In each house lives a person of a different nationality.
3. Each of these five homeowners drink a different kind of beverage, smoke different brands of cigars, and keep different pets.

The question the reader should try to answer is, "Who owns the fish?"

Einstein offers the following clues:

1. The Brit lives in the red house.
2. The Swede keeps dogs as pets.
3. The Dane drinks tea.
4. The green house is on the left of the white house.
5. The owner of the green house drinks coffee.
6. The person who smokes Pall Mall rears birds.

7. The owner of the yellow house smokes Dunhill.
8. The man living in the center house drinks milk.
9. The Norwegian lives in the first house.
10. The man who smokes Blends lives next to the one who keeps cats.
11. The man who keeps horses lives next to the man who smokes Dunhill.
12. The man who smokes Blue Master drinks beer.
13. The German smokes Prince.
14. The Norwegian lives next to the blue house.
15. The man who smokes Blends has a neighbor who drinks water.

The answer to the riddle is found later in this book.

Acknowledgments

To give credit where credit is due, the Shingo Institute would need to create a long list of individuals, companies, and universities that the Institute has worked with through the years. The list is far too long to give everyone credit, but the Shingo Institute needs to acknowledge a few specific individuals who were instrumental in the development of this book's material. Those most closely involved during their tenure at the Shingo Institute were Shaun Barker, Max Brown, Randall Cook, Robert Miller, Gerhard Plenert, and Jacob Raymer. The Shingo Institute is appreciative of their contributions for without which the content of this book would not exist.

The Shingo Institute would also like to thank the members of the Shingo Executive Advisory Board and affiliate organizations who provide practical insights and critical feedback. And a special thanks to the Jon M. Huntsman School of Business at Utah State University for providing the Institute with a home and an environment to learn, flourish, and grow.

Finally, the Shingo Institute thanks Gerhard Plenert for compiling this book. Dr. Plenert has lived and worked in factories in North, Central, and South America; Asia; the Middle East; and Europe. He has coauthored articles and books and worked with academics and professionals from as far away as Europe, Japan, and Australia. His experience has provided him a broad exposure to a variety of manufacturing, service, government, and military facilities all over the world.

1

Introduction

Too many organizations are failing to be competitive, not because they cannot solve problems, but because they cannot sustain the solution. They haven't realized that tradition supersedes tools, no matter how good they are. Success requires a sustainable shift in behaviors and culture, and that needs to be driven by a shift in the systems that motivate those behaviors.

Gerhard Plenert

Countless organizations have, at one time or another, begun a "Lean journey" (for a brief description of Lean, please see the end of this introduction) or they have implemented an improvement initiative of some sort. At the foundation of these initiatives are a number of tools that seem to promise exciting new results. While many organizations may initially see significant improvements, far too many of these initiatives meet disappointing ends. Leaders quickly find that Lean tools such as Six Sigma, *jidoka*, SMED, 5S, JIT, quality circles, etc. are not independently capable of effecting lasting change.

Years ago, the Shingo Institute set out on an extended study to determine the difference between short-lived successes and sustainable results. Over time, the Institute noticed a common theme: the difference between successful and unsuccessful effort is centered on the ability of an organization to ingrain into its culture timeless and universal principles rather than rely on the superficial implementation of tools and programs. These findings are confirmed time and again by nearly three decades of assessing organizational culture and performance as part of the Shingo Prize process. Since 1988, Shingo examiners have witnessed first-hand how quickly tool-based organizations decline in their ability to sustain results. On the other hand, organizations that anchor their improvement initiatives to principles experience significantly different results. This is because principles help people understand the "why" behind the "how" and the "what."

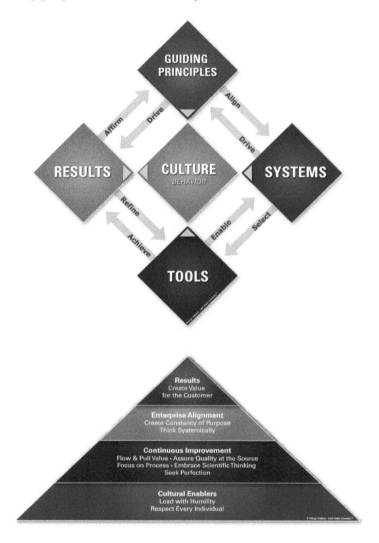

To best illustrate these findings, the Shingo Institute developed the *Shingo Model™*, the accompanying *Shingo Guiding Principles*, and the *Three Insights of Enterprise Excellence™*. The Shingo Institute offers a series of five workshops designed to help participants understand these principles and insights and to help them strive for excellence within their respective organizations. This book, *Discover Excellence: An Overview of the Shingo Model and Its Guiding Principles*, is an introduction to these five workshops.

DISCOVER EXCELLENCE (prerequisite)
Behaviors that lead to enterprise excellence

CULTURAL ENABLERS
Behaviors that enable a culture of respect and humility

CONTINUOUS IMPROVEMENT
Behaviors that improve a continuous flow of value

ENTERPRISE ALIGNMENT & RESULTS
Behaviors that align people, systems, and strategy

BUILD EXCELLENCE
Driving strategy to execution

Here is a description of the first of five workshops offered in the Shingo Institute educational series.

DISCOVER EXCELLENCE WORKSHOP

A facility-wide improvement initiative is expensive in terms of both time and money. Perhaps the most disappointing thing about them is that they often end up as temporary measures that may produce early results but are unsustainable in the long run. The unseen cost is that after they see such initiatives come and go, employees begin to see them as futile, temporary annoyances rather than the permanent improvements they are meant to be.

The *Shingo Model* begins with culture informed by operational excellence principles that lead to an understanding of what aligns systems and tools and can set any organization on a path toward enterprise excellence with sustainable continuous improvement.

The *Shingo Model* is not an additional program or another initiative to implement. Instead, it introduces *Shingo Guiding Principles* on which to anchor current initiatives. Ultimately, the *Shingo Model* informs a new way of thinking that creates the capability to consistently deliver ideal results to all stakeholders. This is enterprise excellence—the level of excellence achieved by Shingo Prize recipients.

DISCOVER EXCELLENCE is a foundational, two-day workshop that introduces the *Shingo Model*, the *Shingo Guiding Principles*, and the *Three Insights of Enterprise Excellence*. With active discussions and on-site learning at a host organization, this program is a highly interactive experience. It is designed to make learning meaningful and immediately applicable as participants learn how to release the latent potential in an organization to achieve enterprise excellence. It provides the basic understanding needed in all Shingo workshops; therefore, it is a prerequisite to the CULTURAL ENABLERS, CONTINUOUS IMPROVEMENT, and ENTERPRISE ALIGNMENT & RESULTS workshops, and concludes with the BUILD EXCELLENCE workshop.

DISCOVER participants will:

- Learn and understand the *Shingo Model*.
- Discover the *Three Insights of Enterprise Excellence*.
- Explore how the *Shingo Guiding Principles* inform ideal behaviors that lead to sustainable results.
- Understand the behavioral assessment process using a case study and on-site learning.

The additional four Shingo workshops are described as follows.

CULTURAL ENABLERS WORKSHOP

The CULTURAL ENABLERS workshop is a two-day workshop that integrates classroom and on-site experiences at a host facility to build upon the knowledge and experience gained at the DISCOVER EXCELLENCE workshop. It leads participants deeper into the *Shingo Model* by focusing on the principles identified in the Cultural Enablers dimension:

- Respect Every Individual
- Lead with Humility

Cultural Enablers principles make it possible for people in an organization to engage in the transformation journey, progress in their understanding, and, ultimately, build a culture of enterprise excellence. Enterprise excellence cannot be achieved through top-down directives or piecemeal implementation of tools. It requires a widespread organizational commitment. The CULTURAL ENABLERS workshop helps participants define ideal behaviors and the systems that drive them using behavioral benchmarks.

CONTINUOUS IMPROVEMENT WORKSHOP

The CONTINUOUS IMPROVEMENT workshop is a three-day workshop that integrates classroom and on-site experiences at a host facility to build upon the knowledge and experience gained at the DISCOVER EXCELLENCE workshop. It begins by teaching participants how to clearly define value through the eyes of customers. It continues the discussion about ideal behaviors, fundamental purpose, and behavioral benchmarks as they relate to the principles of Continuous Improvement, and takes participants deeper into the *Shingo Model* by focusing on the principles identified in the Continuous Improvement dimension:

- Seek Perfection
- Embrace Scientific Thinking
- Focus on Process
- Assure Quality at the Source
- Flow & Pull Value

This workshop deepens one's understanding of the relationship between behaviors, systems, and principles and how they drive results.

ENTERPRISE ALIGNMENT & RESULTS WORKSHOP

The ENTERPRISE ALIGNMENT & RESULTS workshop is a two-day workshop that integrates classroom and on-site experiences at a host facility to build upon the knowledge and experience gained at the DISCOVER EXCELLENCE workshop. It takes participants deeper into the *Shingo Model* by focusing on the principles identified in the Enterprise Alignment dimension and the Results dimension:

- Think Systemically
- Create Constancy of Purpose
- Create Value for the Customer

To succeed, organizations must develop management systems that align work and behaviors with principles and direction in ways that are simple, comprehensive, actionable, and standardized. Organizations must get results, and creating value for customers is ultimately accomplished through the effective alignment of every value stream in an organization. The ENTERPRISE ALIGNMENT & RESULTS workshop continues the discussion around defining ideal behaviors and the systems that drive them, understanding fundamental beliefs, and using behavioral benchmarks.

BUILD EXCELLENCE WORKSHOP

The BUILD EXCELLENCE workshop is the two-day capstone workshop that integrates classroom and on-site experiences at a host facility to solidify the knowledge and experience gained from the previous four Shingo workshops. BUILD EXCELLENCE demonstrates the integrated execution of systems that drive behavior toward the ideal as informed by the principles in the *Shingo Model*. The workshop helps to develop a structured approach to execute a cultural transformation. It builds upon a foundation of principles, using tools that already exist within many organizations. It teaches participants how to build systems that drive behavior, which will consistently deliver desired results.

In this final Shingo workshop, participants will:

- Design or create a system, guided by the *Shingo Model*, that changes behaviors to close gaps and drives results closer to organizational goals and purpose.
- Answer the question: "How do I get everyone on board?"
- Build on the principles of enterprise excellence.
- Understand the relationship between behaviors, systems, principles, and how they drive results.

- Learn how key behavioral indicators (KBIs) drive key performance indicators (KPIs), and how this leads to excellent results.
- Use "go and observe" to understand the practical application of the *Shingo Guiding Principles*.

With this understanding of what this book is all about, the reader can now take the first of many steps toward enterprise excellence.

DEFINITION OF LEAN

The term "Lean" was first adopted by authors James P. Womack, Daniel T. Jones, and Daniel Roos in *The Machine That Changed the World: The Story of Lean Production*. They describe Lean as manufacturing systems that are based on the principles employed in the Toyota Production System (TPS). Quoting them,

> Lean ... is 'lean' because it uses less of everything compared with mass production—half the human effort in the factory, half the manufacturing space, half the investment in tools, half the engineering hours to develop a new product in half the time. Also, it requires keeping far less than half the inventory on site, results in many fewer defects, and produces a greater and ever growing variety of products.[*]

Lean's philosophy has evolved through numerous iterations. It stresses the maximization of customer value while simultaneously minimizing waste. Lean's goal is the creation of increased value for customers while simultaneously utilizing fewer resources. To accomplish this, Lean utilized a plethora of tools (over 100) to optimize the flow of products and services throughout an entire value stream as they horizontally flow through an organization. However, Lean does not capture the focus on cultural shift, which was a necessary part of the original TPS, and which the Shingo Institute attempts to restore. The Shingo Institute uses Lean in the meaning that was intended when it was first coined by Womack, Jones, and Roos.

---

[*] Womack, J. P. et al. *The Machine That Changed the World: The Story of Lean Production—Toyota's Secret Weapon in the Global Car Wars That Is Now Revolutionizing World Industry*. New York, NY: Simon & Schuster, Inc., 1990, p. 14.

Section I

Overview of the *Shingo Model*

2

Why Are We Here?

Are you too busy for improvement? Frequently, I am rebuffed by people who say they are too busy and have no time for such activities. I make it a point to respond by telling people, look, you'll stop being busy either when you die or when the company goes bankrupt.*

Shigeo Shingo
Namesake of Shingo Institute

HOW TO DEFINE ENTERPRISE EXCELLENCE?

The road to perfection is one that is constantly shifting, and therefore requires numerous course corrections as one travels along. One should ask, "Is the enterprise aligned with the type of culture that promotes excellence? Is the enterprise striving for excellence?"

To begin the journey, it is best to first define enterprise excellence. The Shingo Institute asked some of the adherents to the Shingo process and some Shingo Prize recipients for their personal definitions of enterprise excellence and here is what they said:

> Enterprise excellence to me for the Air Force was about trying to instill purposeful change to mitigate the root cause of performance problems. Performance problems being defined as those things that stopped us from advancing the mission.
>
> Being able to mitigate the root cause of those performance problems, we wanted to embed that in the DNA of every airman so they could wake up thinking about how to improve their job, how to make their mission set better,

---

* Shingo, S. AZQuotes.com. Wind and Fly LTD, 2017. Available at www.azquotes.com/quote/731325.

and how to improve teamwork and integration capabilities so that achieving the stated mission and objectives was just a natural course of things.

Brou Gautier
President, Total Systems Development (TSD)
Lexington, Kentucky

For us, enterprise excellence is really about how everybody, every day they come to work. And improving the work is the work. That it is not just about doing the processes. It's about how you're going to improve today. We follow the Shingo Model *in that respect because we're a company where it's not just about results, but how you get results.*

Seán Kelly
Global Business Excellence Program Manager, Abbott Diagnostics
Longford, Ireland

What's truly in my view of excellence is all about engaging the whole work-force and making a better company, and we've managed to do that over the years. A key element of that has been employee engagement and designing our approach to the continuous approach based on getting people in there and getting them to take responsibility for the processes they are a part of.
    When you do that over a number of years you become excellent.

Christian Houborg
Vice President, Lundbeck Pharma A/S
Valby, Denmark

For me, excellence enterprise is related to creating a great environment for the people, looking for the same objectives, and just having fun everywhere.

Maria Suarez
Plant Manager, Rexam Beverage Can Americas (acquired by Ball Corporation), Santiago de Querétaro, Querétaro, Mexico

Enterprise excellence is a status in which an organization has achieved not only financial or market results, but transcended to a different level where respectful people, culture, and principles are key factors in a strategy to sustain growth over generations.

Fernando Ramirez Garza
Quality Director, Deacero
Monterrey, Nuevo León, Mexico

It would be that we understand what our customer needs are. We're working with them to develop solutions to better meet those needs every day, day in and day out.

And we're relevant in the eyes of our customer and we continue to challenge what that relevancy looks like for us.

Elaine Richardson
Senior Advisor, Operation Excellence, Export Development Canada
Ottawa, Ontario, Canada

Operational excellence is an aspirational state where we're trying to optimize the value for our customers through eliminating waste and engaging our employee base, so we can have every employee delivering at their full potential and delivering as much value to our customers as possible.

Melanie Foley
Executive VP, Chief Talent and Enterprise Services Officer, Liberty Mutual
Boston, Massachusetts

I think being part of an enterprise excellence environment is being never satisfied. Constantly moving forward, constantly thinking.

I remember reading one time about this top company, and what kept them the top company was believing they were #2. If we start thinking that we're #1, someone else is going to creep up behind us. I thought that was a really powerful way of looking at it.

Noel Hennessey
Continuous Improvement Director, Lake Region Medical Ltd.
Galway, Ireland

You are part of an excellent organization. An organization where it's safe to challenge everything. The people you work with respect you. Your leaders and coworkers are humble. It's a learning organization.

Scott Powell
Director, Operation Excellence, Export Development Canada
Ottawa, Ontario, Canada

What's critical about enterprise excellence, at least with the way that's defined by Shingo, it does talk about tools, which is fantastic, and systems how all these things hang together to be effective.

I think what's really valuable is the focus on culture and behaviors. For our journey, that was the most important thing: to win the hearts and minds of our people first.

Michelle Lue-Reid
GM Group Productivity, Commonwealth Bank of Australia (CBA)
Sydney, New South Wales, Australia

Being a part of an enterprise excellence operation, it introduces a totally different culture. It enhances our culture because it brings focus on the customer and the customer experience.

Wendell Haywood
AVP, Life & Health Claims, State Farm Insurance
Bloomington, Illinois

The answer to the question, "What is enterprise excellence to you and your organization?" will be different for everyone. No two definitions are the same. But here are some of the key points that the Shingo Institute learned from these quotes. Enterprise excellence is:

Trying to instill purposeful change to mitigate the root cause of performance problems.

Improving the work is the work. It is not just about doing the processes.

Engaging the whole workforce and making a better company.

Creating a great environment for the people, looking for the same objectives, and having fun everywhere.

A status in which an organization has achieved not only financial or market results, but transcended to a different level where respectful people, culture, and principles are key factors in a strategy to sustain growth over generations.

Understanding what customer needs are.

Trying to optimize the value for customers.

Being never satisfied. Constantly moving forward, constantly thinking.

An organization where it's safe to challenge everything.

A focus on culture and behaviors.

One that enhances culture because it brings focus on the customer.

Traditionally, when one discusses organizational excellence, they refer to "operational excellence;" however, "enterprise excellence" requires a much broader perspective. It is important to look at the enterprise as a whole. The Shingo Institute is moving away from the term "operational excellence," since that would eliminate consideration of many other key departments such as HR, finance, etc. Instead, it wants to focus on the enterprise as an integrated entity, including all internal organizations as well as the full value chain (everyone from the supplier's supplier to the customer's customer). Enterprise excellence involves all employees, departments, customers, contractors, and vendors. It is the pursuit of excellence of the entire extended organization. To be achieved, a culture of excellence must be embraced by all. It must be continuously and purposefully pursued. Driving toward excellence cannot be a movement that becomes the "fad of the month" for leadership. It has to be an integrated and united effort to succeed. These are the first steps in a long journey toward perfection.

In this book, the reader will discover how many of the world's best-run organizations have achieved, or are on their way toward achieving, their goal of enterprise excellence. In this journey toward excellence, continuously improving enterprises have traditionally focused on results, primarily KPIs such as sales, revenue, stock price, or net profit. They achieve occasional spot improvement in KPI results using a tool-based approach. This includes tools coming out of the TPS such as value stream mapping (VSM), rapid improvement events (RIEs), and 5S. In addition, they utilize continuous improvement strategies such as Lean, Six Sigma (6σ), Theory of Constraints, Green, and many more. As stand-alone, these tools can provide spot results. But to achieve an environment of long-term, sustainable continuous improvement throughout the enterprise, more than just tools, will be required.

Does this sound familiar? Is the reader's organization an occasional user of tools to fix problems? If so, then here are a few searching questions:

1. Do the tools provide sustainable results or are they just something used once a year and then forgotten the rest of the year (from the perspective of employees)?
2. Do improvements seem to go away as employees refocus on tradition?
3. Are these tools primarily used to solve problems (failure points)? Or should one instead look for opportunities that stretch organizations in new ways that may circumvent and avoid the problems?
4. Do tools improve culture within the organization? For example, do tools cause employees to look for continuous improvement opportunities all year long? Or is a continuous improvement exercise considered as a disruption to the normal work flow?
5. Are the improvement events strategically aligned? Does anyone look at the enterprise strategy to make sure there is a strategic benefit for going through this improvement exercise? Or is the driver strictly about "problem solving"?
6. Are continuous improvement tools used when the reader is involved in long-term strategic thinking, new product introductions, marketing, engineering, and the like?

Returning to the question of excellence, how does one define sustainable enterprise excellence (success) for the enterprise? What if the business took a slightly different perspective on how to achieve enterprise excellence? Instead of focusing solely on KPI results, what if businesses instead considered behavioral elements like doing the right things before doing things right? What if businesses focused on the relationships with and behaviors of customers, quality, employees, suppliers, etc. instead of only KPIs? This is not to say that KPIs are not important. The Shingo Institute suggests that sustainable KPI improvement requires a cultural shift, or the improvement is only temporary.

In their search for improvement, most companies focus on tools that they use to hopefully meet their goals. And the tools do offer them spot improvements. The variety of tools along with the variety of solutions available is almost endless. Why then do enterprises struggle in reaching their goals, sustaining their progress, and achieving enterprise excellence? First look at what has traditionally been their focus. As shown in Figure 2.1, the bridge toward sustainability requires more than just a pillar based on tools. It is important to construct the second supporting pillar for the bridge to be useful, and that's the primary topic of this book.

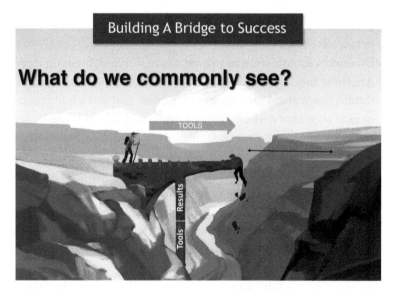

FIGURE 2.1
Bridge to Operational Excellence.

Even though these tools provide impressive results, the achievements are often fleeting and not long lasting. What is lacking is sustained superior performance, a sustained culture of excellence and innovation, and a sustained environment for social and ecological leadership. There is a need for long-term sustainability to really make progress in a journey to enterprise excellence. It is important to close the gap in the bridge by building a sustainable structure that supports a shift in culture.

The most widely used model in the world for this journey to enterprise excellence is called the *Shingo Model.* It is based on 10 guiding principles that are grouped into four dimensions. The principles are timeless and universal. They apply to all cultures and nationalities and do not change over time. These principles provide a solid foundation for developing a roadmap to excellence. But before the reader can dive deeply into the Model itself, it is time to take a short journey through its history.

A LITTLE HISTORY

Very few individuals have contributed as much to the development of the ideas of TPS, TQM (total quality management), JIT (just-in-time), Six Sigma (6σ), and Lean as did Shigeo Shingo. Over the course of his life, Dr. Shingo wrote and published 17 books, 8 of which have been translated from Japanese to English. Many years before these concepts became popular in the Western world, Dr. Shingo wrote about the ideas of ensuring quality at the source, flowing value to customers, working with zero inventories, rapidly setting up machines through the system of "single-minute exchange of dies" (SMED), and going to the actual workplace to grasp the true situation there (known in the West as going to gemba but Dr. Shingo preferred to call it "go and observe"). He worked extensively as a consultant with Toyota executives, especially Taiichi Ohno, who helped him to apply his understanding of these concepts to the real world. Dr. Shingo was instrumental in designing and implementing the very successful and internationally recognized TPS. The foundational concepts or principles from this production system can be applied to a multitude of organizations. They have now spread across all industries including healthcare, insurance, government, military, logistics, and the list goes on and on. Any organization with a process or system can incorporate TPS principles and find benefit from them.

Shigeo Shingo

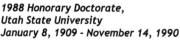

1988 Honorary Doctorate,
Utah State University
January 8, 1909 - November 14, 1990

Always on the leading edge of new creativity and continuous improvement, Dr. Shingo envisioned a collaboration with an organization that would further his life's work through research, practical yet rigorous education, and a program for recognizing the best in enterprise excellence throughout the world. In 1988, Shingo received his honorary doctorate of management from Utah State University and, later that year, his ambitions were realized when the Shingo Prize was created and administered by the university.

THE SHINGO PRIZE

The Shingo Prize has become the world's highest standard for enterprise excellence. As an effective way to benchmark progress toward enterprise excellence, organizations throughout the world may apply and challenge for this Prize. Recipients receiving this recognition fall into three categories:

The Shingo Prize is awarded to organizations that have robust key systems driving behavior close to ideal, as informed by the principles of operational excellence, and supported by strong key performance indicator and key behavioral indicator trends and levels. Shingo Prize recipients show the greatest potential for sustainability as measured by the frequency, intensity, duration, scope, and role of the behaviors evident in the organizational culture.

The Shingo Silver Medallion is awarded to organizations that are well along the transformation path and heading in an appropriate direction as it relates to principles, systems, tools, and results. Behaviors and measures show results from a focus on key systems. Significant progress has been made with respect to frequency, intensity, duration, scope, and role of the behaviors evident in the organizational culture.

The Shingo Bronze Medallion is awarded to organizations that are at the developmental stage as it relates to principles, systems, tools, and results. Behaviors and measures are identified and the organization is working toward stability in both. Progress is made with respect to frequency, intensity, duration, scope, and role of the behaviors evident in the organizational culture.

Most organizations do not wait until they believe they might qualify for the Shingo Prize to challenge. Rather, they use this progression as a way to guide their journey of continuous improvement. Many organizations do not intend to ever challenge for the Prize, but use the *Shingo Model* and the Prize Assessment process to measure themselves as they work toward the highest standard of excellence in the world. They use these guidelines as a tool to direct them, to which they aspire, and to which they hold themselves accountable.

The desire to improve seems to be instinctive. For any organization to be successful in the long term, it must engage in a relentless quest to make things better. In fact, if an organization is to survive and thrive, leaders must motivate their organizations to be on a continuous pursuit of perfection. Although it is fundamentally impossible to achieve perfection, the pursuit of it can bring out the very best in every organization and in every person. Improvement is hard work! It requires great leaders, smart managers, and empowered associates. Sustainable improvement cannot be delegated down nor organized into a "flavor-of-the-month" program or initiative. It requires a fundamental cultural shift and that is what the Shingo methodology drives one toward.

Improvement requires more than the application of a new tool or a leader's charismatic personality. Excellent sustainable results require the transformation of a culture to one where every single person is engaged every day in making small, and from time-to-time large, changes. In reality, every organization is naturally in some state of transformation. The critical question is, "Into what is the organization being transformed?"

Since the establishment of the Shingo Prize at Utah State University in 1988, the Shingo Institute has assessed organizations in various industries around the world as they've challenged for a Shingo award. The Institute has seen firsthand how some organizations have been able to sustain their improvement results, while far too many have experienced a decline. Years ago, the Institute discovered a clear theme, "Sustainable results depend upon the degree to which an organization's culture is aligned to specific, guiding principles rather than depending solely on tools, programs, or initiatives."

To graphically illustrate these findings, the Institute developed the *Shingo Model* (discussed in detail in Chapter 6), including the *Shingo Guiding Principles* (Chapter 7) and the *Three Insights of Enterprise Excellence* (Chapter 5). The *Shingo Model* provides a powerful framework that will guide one in transforming their organization's culture and achieving ideal results.

> *It is no longer good enough to be good, we want to be great!* The Shingo Guiding Principles, *founded in logic and built over time, have assisted us*

in moving further toward operational excellence. As we adapted our site to align with these principles, we found ourselves doing things that just make sense. This is testament to the power of the principle-led approach.

Pat Kealy
Business Excellence Manager, Abbott Vascular in Clonmel, Ireland
2014 Shingo Prize Recipient

A list of the Shingo recipients through this book's printing is included in the Appendix. A complete and updated list is also available on the Shingo Institute website at http://shingo.org/awards.

THE SHINGO INSTITUTE

The Shingo Institute was founded to help companies across the globe strive to understand, embrace, and practice those successful Shingo concepts that date back to the original development of TPS. The Shingo Institute is housed in the Jon M. Huntsman School of Business at Utah State University. It is the home of the Shingo Prize, an international award for enterprise excellence; Shingo conferences and summits; Shingo workshops including the one described in this book; and the Shingo Insight™ survey tool.

In 2008, after an intensive study regarding the necessary components of creating a company culture that is able to sustain improvements and consistently drive results, the Shingo Institute realized it needed to "raise the bar." The Institute's attention shifted from an emphasis on a tool and programmatic assessment primarily focused on operational excellence, toward a complete assessment of the organization's culture as a whole, including all the interactions between the various functions of the organization.

While the Shingo Prize remains an integral part of the Shingo Institute, the scope has expanded to include various educational offerings, a focus on research, and a growing international network of Shingo Institute Licensed Affiliates. Volunteer Shingo examiners, who are international experts in all aspects of enterprise excellence, now focus on determining the degree to which the *Shingo Guiding Principles* are evident in the behavior of every employee. They observe behavior and determine the frequency, duration, intensity, and scope of the desired principle-based behaviors. They observe the degree to which leaders are focused on principles and culture, and the degree to which managers are focused on aligning systems to drive ideal behaviors at all levels.

The focus at the Shingo Institute is unique in the world, and is the most rigorous way to determine if an organization is fundamentally improving over the long term or just going through the motions of another flavor-of-the-month initiative. As with the philosophy of the *Shingo Model*, the Shingo Prize has continued to evolve and strives to create its own brand of excellence based on the direction of Shigeo Shingo and the TPS, since its creation in 1988. The following year the first Shingo Prize was awarded, and in 1993, in an effort to help companies achieve the Prize, the initial structure of the *Shingo Model* was developed. In 2000, *Business Week* referred to the Shingo Prize as the "Nobel Prize for Manufacturing." A short five years later, the first international Shingo conference was held in Mexico. With this growth came more improvements, and in 2008, the Bronze and Silver Medallion levels were created. Many Shingo Prize recipients are household names: Abbott, Autoliv, Ball, Baxter, Boston Scientific, Denso, Ford Motor Company, Goodyear, John Deere, Johnson Controls, Lycoming Engines, O.C. Tanner Company, Steelcase, and U.S. Synthetic to name a few.

DRIVING TOWARD ENTERPRISE EXCELLENCE

Initially, the famous Shingo Prize focused on tools and systems, and how those tools and systems drive results. The Prize was given out based on these results. But unfortunately, when far too many award-winning companies reverted to their old ways, the Shingo Institute realized there was a big piece missing in its earlier tool-based definition of enterprise excellence. It needed to be principle-based and needed to point toward sustainability.

To create sustainable improvement, the Institute needed to make its own cultural shift. It needed to stop using tools as the foundation for change, and instead embed strategies for continuous improvement focused on a cultural shift within the enterprise. Change no longer is something that happens once a year during a Lean event. Instead, organizations need to constantly look for improvement opportunities. As Masaaki Imai says, the definition of kaizen is not just "continuous improvement" but rather it should be "every day improvement, everybody improvement, and everywhere improvement."

In a quest to share their knowledge and insight with the rest of the world, the Shingo Institute developed a model that is the primary subject of a series of workshops and books. These materials were developed specifically to share the knowledge of how to create a sustainable cultural shift, which will ultimately lead to enterprise excellence. This set of workshops was created in this drive toward helping organizations achieve enterprise excellence.

DISCOVER EXCELLENCE (prerequisite)
Behaviors that lead to enterprise excellence

CULTURAL ENABLERS
Behaviors that enable a culture of respect and humility

CONTINUOUS IMPROVEMENT
Behaviors that improve a continuous flow of value

ENTERPRISE ALIGNMENT & RESULTS
Behaviors that align people, systems, and strategy

BUILD EXCELLENCE
Driving strategy to execution

The Shingo Institute offers five distinct workshops to build enterprise excellence. The first workshop, DISCOVER EXCELLENCE, is a foundational, two-day workshop that introduces the *Shingo Model*, the *Shingo Guiding Principles*, and the *Three Insights to Enterprise Excellence*. With active discussions and on-site learning at a host organization, this program is a highly interactive experience. It is designed to make learning meaningful and immediately applicable as participants learn how to release the latent potential in an organization to achieve enterprise excellence. It provides the basic understanding needed in all Shingo workshops; therefore, it is a prerequisite to the CULTURAL ENABLERS, CONTINUOUS IMPROVEMENT, and ENTERPRISE ALIGNMENT & RESULTS workshops, and concludes with the BUILD EXCELLENCE workshop.

The second workshop, CULTURAL ENABLERS, is a two-day workshop that integrates classroom and on-site experiences at a host facility to build upon the knowledge and experience gained at the DISCOVER EXCELLENCE workshop. It leads participants deeper into the *Shingo Model* by focusing on the principles identified in the Cultural Enablers dimension: Respect Every Individual and Lead with Humility. Cultural Enablers principles make it possible for people in an organization to engage in the transformation journey, progress in their understanding, and, ultimately, build a culture of enterprise excellence. Enterprise excellence cannot be achieved through top-down directives or piecemeal implementation of tools. It requires a widespread organizational commitment. The CULTURAL ENABLERS workshop helps participants define ideal behaviors and the systems that drive them using behavioral benchmarks.

CONTINUOUS IMPROVEMENT, the Shingo Institute's third workshop, is a three-day workshop that integrates classroom and on-site experiences at a host facility to build upon the knowledge and experience gained at the DISCOVER EXCELLENCE workshop. It begins by teaching participants how to clearly define value through the eyes of customers. It continues the discussion about ideal behaviors, fundamental purpose, and behavioral benchmarks as they relate to the principles of Continuous Improvement, and takes participants deeper into the *Shingo Model* by focusing on the principles identified in the Continuous Improvement dimension: Seek Perfection, Flow & Pull Value, Assure Quality at the Source, Focus on Process, and Embrace Scientific Thinking. This workshop deepens one's understanding of the relationship between behaviors, systems, and principles and how they drive results.

The fourth workshop, ENTERPRISE ALIGNMENT & RESULTS, is a two-day workshop that integrates classroom and on-site experiences at a host facility to build upon the knowledge and experience gained at the DISCOVER EXCELLENCE workshop. It takes participants deeper into the *Shingo Model* by focusing on the principles identified in the Enterprise Alignment dimension and the Results dimension: Think Systemically, Create Constancy of Purpose, and Create Value for the Customer. To succeed, organizations must develop management systems that align work and behaviors with principles and direction in ways that are simple, comprehensive, actionable, and standardized. Organizations must get results, and creating value for customers is ultimately accomplished through the effective alignment of every value stream in an organization. The ENTERPRISE ALIGNMENT & RESULTS workshop continues the discussion around defining ideal behaviors and the systems that drive them, understanding fundamental beliefs, and using behavioral benchmarks.

The final, two-day capstone workshop, BUILD EXCELLENCE, integrates classroom and on-site experiences at a host facility to solidify the knowledge and experience gained from the previous four Shingo workshops. BUILD EXCELLENCE demonstrates the integrated execution of systems that drive behavior toward the ideal as informed by the principles in the *Shingo Model*. The workshop helps to develop a structured approach to execute a cultural transformation. It builds upon a foundation of principles, using tools that already exist within many organizations. It teaches participants how to build systems that drive behavior, which will consistently deliver desired results.

As there is far more to learn, discuss, and experience than one could ever manage to express and include in written form, the Shingo Institute invites the reader to further their education by exploring additional avenues of education. Beyond the workshops, the Institute also has a wide array of Licensed Affiliates who can bring years of experience and expertise to clients. They are located throughout the world and are available to support those on their journey, in their own country, and in their own language. To learn more about each of our affiliates, the reader should visit shingo.org/affiliates and contact the affiliate that best matches their needs. In addition, the Shingo Institute offers private workshops for organizations upon request. These in-house workshops are tailored so that learning is specifically centered and focused on an individual organization's culture. Licensed Affiliates are often available to facilitate workshops on-site and, with years of consulting experience in various industries, can assist those with specific challenges.

DISCOVER EXCELLENCE

This book, *Discover Excellence*, is a supplement to the first workshop in the Shingo educational series. The DISCOVER EXCELLENCE workshop is a two-day event with the first day in lecture and the second day practicing the learning in a team-based "go and observe" experience. The following are the DISCOVER EXCELLENCE workshop's primary learning objectives:

- Understand the principles of enterprise excellence
- Learn the key insights of ideal behaviors
- Understand the relationship between behaviors, systems, and principles
- Learn how systems and behaviors drive results
- Learn how KBIs drive KPIs and how this leads to excellent results
- Use "go and observe" to understand the practical application of the *Shingo Guiding Principles*

Earlier in this chapter, a variety of people in various industries shared different perspectives on what enterprise excellence means. The Institute is not defining enterprise excellence as only "one way of thinking" or correcting opinions on how these individuals define it for themselves. Rather, the Institute offers another perspective that will help the reader create sustainable cultures of excellence.

At this point, the Shingo Institute invites the reader to reflect back on the various opinions of the individuals interviewed at the start of this chapter and ask the following:

- What was the central message of each individual? What does enterprise excellence mean to them?
- What two or three things did these individuals highlight that are important?
- What are some specific things the reader can take back with them to work?
- What will the reader do differently?

Now that enterprise excellence has been defined, it is time to press forward with a drive to successfully achieve it.

> Be a yardstick of quality. Some people aren't used to an environment where excellence is expected.*

Steve Jobs
Co-founder, former Chairman & CEO, Apple Inc.

---

* Young, J. *Steve Jobs: The Journey is the Reward*, Lynx Books, 1988.

3

Defining Behavior

*Even the greatest idea can become meaningless in the rush to judgment. To gauge an idea as feasible, we must cut our ties to the status quo and find the balance between constructive criticism and judgment. Within that balance we will uncover crucial input to make our dreams a reality.**

Shigeo Shingo

WHY DO ENTERPRISES EXIST?

Why are enterprises created and why do they exist? The answer can be stated in one word, "Results!" The next question should be, "What are the results the enterprise attempts to achieve?" This answer can take numerous forms. For most enterprises, results come in the form of KPIs. In a for-profit enterprise, results are focused on sales, revenue, stock price, profits, return on investment (ROI), return on net assets (RONA), etc. In a not-for-profit enterprise, which includes charities, government entities, military, and so on, results are measured in terms of cost, budget, percentage of cost to budget, etc. Occasionally, but not too often, KPIs take the form of some measure of nonfinancial benefits.

If enterprises exist to achieve results, then it is important to determine the best way to achieve those results. Traditional enterprise performance methodology shows that results come by increasing sales and cutting costs. Unfortunately, tradition leads in directions that can cause one to make significant mistakes. For example, tradition shows that employees are a

---

* Shingo, S. *Kaizen and the Art of Creative Thinking*, Bellingham, WA: Enna Products Corporation and PCS Inc., 2007, p. 169.

cost and therefore need to be minimized. Since employee cost is one of the easiest costs to reduce, since all one has to do is fire someone, it is often the cost that is attacked first. However, cutting costs by cutting employees tends to have a devastatingly negative effect on the organization. It damages morale, and often eliminates the key individuals that are the most critical for the growth and stability of the company, simply because they are the most expensive employees within the organization.

To get results, one needs two things: systems and tools. The tools are the easy part to understand. Tools can be assets, such as machinery, facilities, and equipment; or tools can take the form of procedures and methods, such as value stream mapping, computer schedules, project management PERT or Gantt charts, and so on depending on what type of industry the reader supports. Tools are important and critical in maintaining an organization. One can't exist without tools or they have no purpose. Examples of tools also include laptops, forklifts, salary, a strategic plan, specific awards, health benefit plans, newsletters, bulletin boards, buildings, and on and on. Think about the reader's own organization and try to list the tools that are critical to success. The list tends to become quite long.

The second thing needed to obtain results is systems. Systems can also take numerous forms. Systems include procedures, like how to set up a drill press, or how to process a loan application. Every organization, no matter the size, has a large number of systems. Examples include planning systems, compensation systems, rewards systems, scheduling systems, HR systems, continuous improvement systems, employee development systems, product development systems, inventory systems, customer support systems, sales systems, marketing systems, IT systems, and many

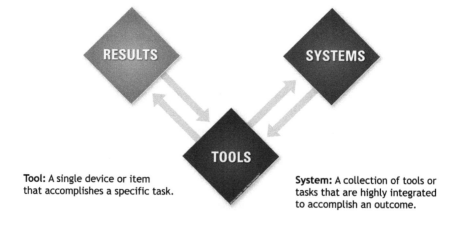

Tool: A single device or item that accomplishes a specific task.

System: A collection of tools or tasks that are highly integrated to accomplish an outcome.

more. The reader should attempt to list all the systems within their organization. Then, after a list is created, take a walk around the organization and quickly realize how many systems the reader has forgotten to list, like quality systems, parking systems, cleaning systems, and on and on.

As enterprises grow, they need to specialize. They create suborganizations (legal, HR, etc.). These in turn start to create their own set of systems such as production systems, quality systems, payroll systems, etc. The result is that because of the complexity of the enterprise's tools and systems one can easily lose control. And why is control important? Because this collection of tools and systems may not directly link to and focus on the strategic direction in which organizations want to go. That's when organizational rigor mortis sets in, and that's when tradition takes priority. When one makes a change, the culture of the organization quickly attempts to restore itself back to its equilibrium point, which is the traditional way of doing things. Change fails, which leads to the famous Einstein quote:

Insanity is continuing to do the same thing and expecting different results.

How does one break the cycle of rigor mortis? How does one get change to stick? The answer is found in culture. To transform an organization, one needs to break free from the current culture that is focused on maintaining tradition. It is important to shift to a new culture focused on continuous improvement. This book will help the reader understand what is necessary for an enterprise to break free.

If one already has systems and tools in place, what else does the reader need? One needs to focus on more than just the system or the tool. One needs to focus on the people behind the systems and the tools. One needs to focus on the enterprise's culture, which is central to all the other pieces. So how does the reader find what their enterprise's culture is composed of? The Shingo Institute administers an award for enterprise excellence, the Shingo Prize, and part of that process is for a team of examiners to physically visit the enterprise being examined. Examiners go there to look at the systems, tools, and culture and to see if these are in harmony with the strategic goals and objectives of the organization. However, what they find is that they can determine the culture of an organization very quickly. Examiners can sense the culture just by walking around the organization and observing the behavior of the people. If they are smiling, willing to make eye contact, eager to talk about what they are doing, and generally seem happy, then the culture is probably a very positive one. However, if the people do not make eye contact, and respond to the examiner's questions with very short answers, basically trying to get rid of the interviewer, then the examiner knows that the culture of continuous improvement does not exist. People are not excited about change. Employees see the change agent as someone introducing a new "fad of the month" and they don't want to get involved.

As an organization grows, it specializes into departments. This results in numerous systems and silos, which are often overlapping and sometimes even conflicting. Each department becomes its own independent culture. These often do not harmonize into an enterprise-wide culture. A silo culture is a culture of competitive conflict. This type of organization needs to rethink its structure and consider more of a value stream culture, where each element fits into an overall value chain and where all the parts interconnect and interface with each other. This is imperative to rebuild the culture of an enterprise, with interconnected parts of a greater whole.

How does one change the culture of an organization? Culture comes from values or beliefs. This does not refer to religious beliefs, but what a person believes about themselves and their fit within the organization. Does the individual believe they have value within their organization? Do they feel their ideas and performance make a difference? Or do they simply do "whatever the boss says" so they can walk home with a paycheck at the end of the day?

What do we need to do to drive results?

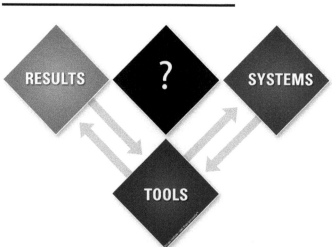

Culture: Values, beliefs, behaviors.
Behavior: Can be observed, described and recorded.

One shifts an organization's culture by shifting the values and beliefs of those within the organization. This is done by building systems that reward people for behaving in a given manner. This process begins by recognizing and understanding the values and beliefs held by an organization - the current state. It can be accomplished by assessing behaviors within an organization–how people act and behave–based on what they truly value and believe. Keep in mind that the people being assessed are the same individuals that build the systems and tools the organization uses. They behave to a large extent based on what is currently rewarded. It becomes critical to focus on behaviors because these help recognize the values and beliefs of both the organization and the individuals within the organization.

DEFINING BEHAVIOR?

A behavior is described as the way one acts or conducts oneself, especially toward others. A behavior is observable, describable, and recordable. It can be shared. It is not an attitude or emotion, or even a piece of knowledge or information. Behavior is defined in the act itself, and it is guided by beliefs and values. It is something that is demonstrated by individuals and observable by onlookers. Given these definitions and explanations, which of the following would the reader consider to be a behavior?

1. Associates know the vision and mission.
2. Associates understand the vision and mission.
3. Associates talk about the vision and mission at the beginning of each meeting.
4. Managers are confused with what is a priority.
5. Managers come in early and stay late every day.
6. Managers are angry with all the special projects.

It's not as easy as it looks. But the answers, of course, are #3 (talk) and #5 (come in). They are both actions. Why were these two items behaviors while the other selections were not? It simply comes down to demonstrating one's principles through an act. The first (#3) is a discussion involving a corporation's vision and overall mission. The second example (#5) similarly involved an action. Coming in early and staying in late is a behavioral trait because it requires physical effort to accomplish it. Without action, confusion and anger fall out as a state of mind or feeling. They are emotional responses to external stimuli.

Taking the definition to the next level, ideal behaviors are defined as any action that creates outcomes that are focused on the strategic objectives of an organization and that are both sustainable and excellent.

At this point, take a closer look at the linkages between behavior and culture. Ask questions like, "Can we predict behavior?" or "Can we modify behaviors if they are not desirable or ideal?" The answer to the first question, "Can we predict behavior?" is, "Yes!" How does one accomplish this? By understanding the set of guiding principles of the individuals one is considering. Focus on the understanding to predict how they will respond to a variety of external stimuli. The reader should take a little test to see how good they could predict behavior under a certain set of specific situations. Consider each of the following situations and ask what types of behavior to expect from employees if the organization's culture is based on or promotes the following:

- The power hungry
- Fear
- Survival
- Firefighting
- Humility
- Innovation
- Respect
- Trust
- Collaboration

No organization exhibits only one of these characteristics. Often there is a blend of several characteristics. Or there may be isolated areas where the culture is more of one type than another. One set of behaviors often creates another second set of behaviors. For example, if the organization promotes firefighting, and rewards individuals who are good at firefighting rather than those who support stability and consistency, the reader will discover that their organization also supports a survival mentality. The reader will find that employees actually start fires, just so they can be the heroes that put the fire out. This in turn destroys trust, and on and on with a chain reaction of problems. Here are some potential answers to the previous list (this list should not be considered to be all inclusive and the reader may come up with ideas that are even better):

- The power hungry? – not a team player, lacks trust, withholds information and only shares it selectively/plays games.
- Fear? – takes shortcuts, lacks respect for others as observed by not talking about or sharing information, breaks safety rules, gets results done at all costs potentially bypassing safety or quality.
- Survival? – highly focused, erratic, dictatorial, fights, argues, breaks rules, doesn't share/collaborate, maximizes the individual's own position over the needs of the group.
- Firefighting? – frantic and pressured for time, rewards and gratification go to the "fix it" person who becomes the hero rather than offering rewards for stability, incentivized to become firefighters because of the reward system, minimizes the use of standard work, doesn't share the fixes, increases tribal knowledge.
- Humility? – cooperates, accepts responsibility, listens/understands and takes action on what is heard, values the ideas of others, willing to change when learning something new.
- Innovation? – performs structured experiments, shares ideas, collects direct and timely feedback from customers, problem-solves around customer issues, applies problem-solving and experimentation methods, visits customers to see how product is used.
- Respect? – listens, timely, teamwork, willing to learn, values safety for all, doesn't rank ideas based on position.
- Trust? – shares information, capable to learn, takes ownership of problems, talks straight, challenges status quo, keeps commitments, makes decisions to benefit the whole.
- Collaboration? – listens to all areas of the organization, seeks input from all, treats input equally, works with others to solve problems, considers impact to the whole, makes sacrifices for the betterment of the whole.

*I strongly believe that you can't win in the marketplace unless you win first in the workplace. If you don't have a winning culture inside, it's hard to compete in the very tough world outside.**

Doug Conant
Former CEO, Campbell Soup & Chairman, Avon Products

---

* Conant, D. Saving Campbell Soup Company. February 11, 2010. Available at http://www.gallup. com/businessjournal/125687/saving-campbell-soup-company.aspx.

Doug Conant is remembered for his dedication to the principles and culture within his organization. He wrote over 30,000 handwritten thank-you notes to various individuals throughout his enterprise during his tenure at Campbell Soup. He is known for walking throughout the production floor and talking to his team members. What did this communicate to his team? How important was culture to him when he prioritized writing these notes every morning? He recognized individual behavior. Unfortunately, many do not typically recognize and appreciate individuals enough for their behavior. What does this mean to the reader? How does this fit within the organization? Specific to the quote, ask, "What has been your experience with bad cultures and their ability (or inability) to win in the marketplace?"

CONTINUING ON THE JOURNEY TO ENTERPRISE EXCELLENCE?

At this point, the reader should understand that an organization's culture is made up of all of the behaviors evident in the organization. An organization's culture drives certain behaviors, both good ones and poor ones, and those behaviors become either stronger or weaker based on how they are rewarded by the culture.

The reader should also have a better understanding of why tools alone cannot sustain a culture of operational excellence. Although tool use alone may successfully deliver good results, continued use of tools alone will not continue to deliver positive results indefinitely, and will not build a culture of enterprise excellence. For example, one can improve a key performance indicator (KPI) of profitability by cutting costs (firing people), but this behavior can also result in sabotaging future growth, poor quality, and poor morale. A traditional focus that is on results only, generated by tools, is not sustainable. A tools-focused organization is characterized as having "events" after which the team goes away until there is another "event."

In contrast, a sustainable culture of enterprise excellence searches for continuous improvement opportunities in everything the organization does every day. In such cultures, all processes are routinely examined, with a critical eye toward improvement. In a sustainable culture of operational excellence, improvement is part of everyone's work, it is not something that is in addition to their work: sustainable delivery of positive results is the "holy grail" of enterprise excellence.

Another way to understand a cultural shift is to look at it through the measures used. As stressed earlier, tools alone give temporary results, as measured by KPIs. In many cases, KPI improvements are realized shortly after a tool is introduced. Lacking the necessary supporting culture, however, it is not unusual for the improvements to erode over time and for performance to return to pre-tool introduction levels. However, a cultural shift can build sustainability because one changes how to get results or cultural behavior. Key behavioral indicators (KBIs) are measures closest to the behavior and, in most cases, measures of behavior itself. KBIs should be the most closely correlated behavioral indicators of an available KPI. They are the behavioral evidence of what is being done (behavior) to achieve KPI goals and objectives. Put another way, KPIs are measures of *what* is achieved; KBIs are measures of *how* they are achieved. The reader can ask, "Does one achieve results through good sustainable behavior, meaning behavior that has the capability to continue to improve results over a long period of time, or through bad behavior which cannot do the same?"

Leaders and managers are very familiar with KPIs. KBIs on the other hand may not be as commonly used or as well understood. Leaders and managers do look for indicators every day that will give them an idea of how KPIs will be impacted. Consider the following simple example intended to illustrate this point. A leader looks around their sales organization and notices: a lack of urgency among sales people as measured by low call volumes (behavioral indicator), a decrease in visits being made to customers (behavioral indicator), high absenteeism (behavioral indicator), people leaving for home early (behavioral indicator), people not talking about the great sales they made (behavioral indicator), and the best sales person talking about how slow things are (behavioral indicator). Given these observed behaviors, the leader can reasonably predict that results as measured by sales (KPI) may not be as good as desired that month. In this case the KBI or KBIs would be the behavioral indicators that most closely correlate with performance in the KPI.

A sustainable culture of enterprise excellence measures its performance in KBIs (how they achieve desired results), as well as KPIs (the results they have achieved). The ability to change or shift behavior in the organization's culture as it relates to KBIs increases the capability of the organization to achieve better performance, as reflected by its KPIs.

At this point, the reader is ready to jump into the next section of the book to study specifically what a culture should look like based on enterprise excellence principles and ideal behaviors. The reader will understand what behaviors that make up an organization's culture need to look like to achieve the desired sustainable results discussed at the beginning of this book.

Thinking is to man what flying is to birds. Don't follow the example of the chicken when you could be a lark.

Albert Einstein
Theoretical Physicist

---

* Pickles, K. 'Thinking is to man what flying is to birds': Unseen letter from Albert Einstein offering advice to students is rediscovered after 65 years. May 23, 2015. Available at http://www.dailymail.co.uk/news/article-3093810.

Section II

Culture

4

Characteristics of Culture

"Know-how" alone isn't enough! You need to "know-why"! All too often, people visit other plants only to copy their tools and methods.*

Shigeo Shingo

WHAT IS CULTURE?

Culture is one of those words that everyone thinks they understand, but everyone defines it differently. First, here is a definition to get everyone on the same page. Drawing on Chapter 3, the reader learned that culture is a collection of values, beliefs, and behaviors held by an organization and the individuals within the organization. The reader also learned that culture manifests itself in the behaviors exhibited by the organization through its leaders, managers, and associates.

The purpose of this chapter is to define both the current culture of the reader's organization, and the ideal culture that the organization would like to see. This is harder than the organization might think. The current culture needs to be defined using terms such as the following:

- Aggressive versus cooperative employees
- Value stream based versus siloed
- Continuous improvement versus indifferent
- People development oriented versus employees as a cost
- Humble management versus authoritarian management

---

* Shingo, S. *Non-Stock Production: The Shingo System of Continuous Improvement.* Portland, OR: Productivity Press, 1988.

- Environmental and safety oriented versus indifferent/business cost oriented
- Positive motivation/recognition oriented versus demanding performance
- Problem solving versus problem ignoring
- Management on the floor versus never see management
- Strategically focused versus no idea what the strategy is
- Customer oriented versus no idea how or what one does affects the customer
- Strong internal communication within the organization versus no idea how or what the company is doing
- Happy versus grumpy

This is by no means an all-inclusive list of the characteristics of a company's culture, but it gives the organization a good start. Use this list to describe the two types of enterprise culture for the organization. Define

1. The current state culture, and
2. The ideal state culture.

The reader probably found a large number of gaps between the current and ideal states. Ask, "What are the things one likes about the current culture? What are things one doesn't like and would like to change? How does one develop a plan to transform the unfavorable things about the current culture?" The remainder of this book will look at the significance of these cultural gaps and how the reader can close them.

WHO CREATES THE CULTURE?

In order to create the desired cultural transformation, take a close look at who creates the culture and what role each of the enterprise's employees has in creating that culture. So who is responsible for creating culture? The Shingo Institute arbitrarily breaks the enterprise's employees up into three groups: leaders (CEO and VP levels, corporate executives, board of directors, president—those who have responsibility over a group of managers and associates and provide strategic direction for the organization), managers (middle management—those who have responsibility over a group of associates and have tactical responsibility for the execution of the strategy), and associates (supervisors and shop floor employees—those who work under the direction of a manager and have responsibility for the work that they perform). In general, leaders create the culture. Managers make sure systems (which influence behaviors) are aligned to the culture. And associates are mostly focused on the tools and are influenced by the culture. They learn how to work with, or work around, cultural limitations.

When the reader looks at the model in Figure 4.1, which of these diamonds do leaders typically focus on the most? Of course, the answer is "Results." And to get results, where do leaders spend most of their time? The answer most often is "Tools," because they will receive examples such as training, various programs like Lean or Six Sigma, or initiatives, etc. Drilling deeper, ask, "What else do leaders do?" The response is usually something like, "We try to create better systems to drive improved results." However, if the user would drill deeper by asking, "How does one accomplish that?" the reader will learn they use tools such as building factories, buying equipment, hiring people, writing software, etc. These are all tools that are used to get results. They implement management systems like SAP or Oracle, or staffing systems. These are all directed at getting spot results. Culture is left in the background or forgotten completely. It's there but it is

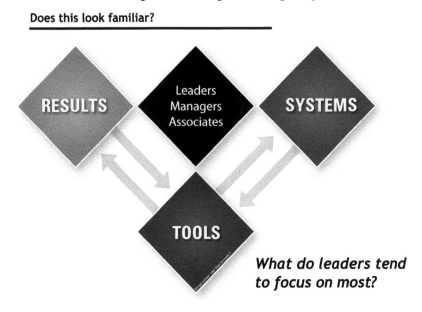

FIGURE 4.1
Typical Framework to Deliver Results.

something else to do when the reader doesn't have more important things to do. Culture is often not the center of what leaders do. But it should be. Most of the time, they look at culture as an HR initiative that sits in the background but is not critical in impacting results. Far too often, this is the reality within many organizations.

BRIDGING THE GAP

There are already systems in organizations, and these systems are motivating employees' behaviors. Tools are focused on results, and this portrays the left-hand side of the bridge. But now it becomes necessary to complete the remainder of the bridge with a focus on sustainable continuous improvement results.

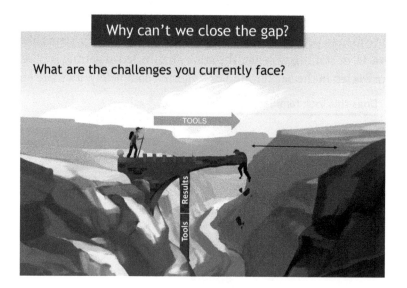

Why is it difficult to close this gap? What challenges do organizations currently face in creating the necessary cultural shift? Some answers might include the following:

- Leaders are constantly changing
- Budget restrictions
- Geopolitical challenges
- Environmental changes

- Bureaucracy
- Competitive landscape that's constantly shifting
- Economic environment
- And many more

Is there a point in the future where any of these things are going to change? Can the reader readily impact any of them? And if the reader could, would that make the gap go away? Probably not. If none of these are going to change, then what does the reader need to do? Can ideal results be achieved within these realities? Of course, it is possible as evidenced by the large list of Shingo recipients listed in the Appendix.

Achieving ideal behaviors and comparing them to current behaviors were already discussed. The reader learned that the current behaviors can be adjusted to move closer to ideal behaviors by adjusting the systems that drive these behaviors. But there is one more point that needs to be made about behaviors, which is that what one assumes is the current set of behaviors is rarely a reality. Find out what the actual current behaviors really are, and not what the assumed current behaviors are. Go out into the work area and observe what is happening. Talk to the associates that actually do the work. With that information, redefine both the current behaviors and the gaps.

One question that often comes up is, "Can one create good results with bad behavior?" and the example that is often cited is Steve Jobs. He was an individual who was very challenging to work for, but he achieved excellent results. And the answer, of course, is, "Yes, one can receive positive results with bad behavior." But the very next question that needs to be asked is, "Are these results sustainable or are they just a flash in the pan?" Sustainability can be defined as a sustained result that maintains itself even if there is a leadership change. Looking at Apple, Jobs' culture didn't sustain after his passing. The aggressiveness of Jobs has been lost, and the culture of Apple is now more challenging to define. And the decline in their results also demonstrates that the loss of Jobs has significantly reduced the organization's performance and effectiveness.

How ideal behaviors achieve ideal results has also been discussed, and a discussion of what those ideal results are will follow. Results, as mentioned in Chapter 2, are the ultimate goal of any enterprise. But what types of results is one looking for? Tradition focuses on KPIs such as revenue, net profit, cost, stock price, ROI, RONA, etc. But rarely are these motivators for the associates on the production floor. They are

FIGURE 4.2
What's Missing?

too abstract for anyone to focus on. These metrics are disconnected from the person working in production. And they are not motivating because the associate doesn't feel that they can influence this number (see Figure 4.2).

KBIs are leading metrics that not only tell where one is headed, but they also help motivate a response from the associates doing the work. KBIs make up the second half of the bridge. They make up the piece that's missing. And they connect to, and are driven by, the culture of the organization. Most important of all, excellent KBI performance results in improved KPI performance.

So, with a discussion about ideal results, one cannot solely look at KPIs and assume they have the complete picture. Ideal results require a foundation of KBIs. Ideal results need to be understandable and measurable by all employees throughout the organization. And these employees need to feel that they can influence and be responsible for the results.

*If you do not know how to ask the right question, you discover nothing.**

W. Edwards Deming
Professor, Author & Consultant

---

* Deming, W. E. *Large List of Quotes by W. Edwards Deming.* Washington, DC: The W. Edwards Deming Institute, 2017. Available at https://blog.deming.org/w-edwards-deming-quotes/large-list-of-quotes-by-w-edwards-deming/.

5

Three Insights of Enterprise Excellence

Improvement usually means doing something that we have never done before.[*]

Shigeo Shingo

INSIGHT

One of the basic requirements behind everything that Toyota does is its undying focus on creativity. The creativity of the individual employee is credited with all their success, including the JIT process, or the TPS methodology, or the various tools such as Lean, Six Sigma, 5S, "go and observe" (gemba), value stream mapping, spaghetti charting, and on and on. Each of these tools blossomed from the creative minds of Toyota employees or Toyota teams.

Shigeo Shingo, who was a consultant to Toyota during these early developmental years, also had his creative juices flowing, and he became the champion of many of the tools that were created. But he wasn't alone. The world of continuous improvement was continually improving, and the years brought a barrage of ideas to the table. Figure 5.1 shows the names of many of these early champions, many of whom are still respected today for their creative insights.

It is difficult to list and give credit to all deserving thought leaders, because there are too many to name throughout the years. There will be many more thought leaders in the future, so readers should not leave with the impression that all things creative have been discovered and that there

---

[*] Shingo, S. AZQuotes.com. Wind and Fly LTD, 2017. Available at www.azquotes.com/quote/731323.

Thought Leaders

Stephen R. Covey

W. Edwards Deming

Rosabeth Moss Kanter

Taiichi Ohno

Henry Ford

Frank & Lillian Gilbreth

SHIGEO SHINGO

Peter Senge

Jim Womack

Joseph M. Juran

Eli Goldratt

Kiichiro Toyoda

Eiji Toyoda

FIGURE 5.1
Thought Leaders.

is nothing new to discover. In a recent conversation, a Toyota executive said to Dr. Plenert:

> *I don't know why people keep copying the Toyota Production System from old books. TPS has changed a lot and continues to change every day. By the time these followers have learned and implemented TPS in their organizations, society will be another 20 to 30 years ahead of them. They will never catch up as long as they keep focusing on copying. If they want to beat us, they need to leapfrog ahead of us with creative new ideas.*

The Shingo Institute has experienced its own insights. Its original Model focused on operational excellence, and the first Shingo Prizes were awarded based strictly on operational performance. Unfortunately, not all these recipients maintained their level of operational performance and some of them reverted to their traditional ways. This left the Shingo Institute and its founders realizing that the *Shingo Model* was incomplete. Enterprise excellence was more than operational excellence. The Model was expanded to what it is today, where it includes enterprise-wide cultural elements. In the end, the expanded Model came a lot closer to the original version Shigeo Shingo and TPS exemplified. However, getting there was a process that resulted in three insights.

So, the big questions were, "How does one sustain excellence?" and "How does one create this cultural shift?" Working with enterprises that had both succeeded and failed at sustaining their results, the Shingo Institute gained the following three insights:

Insight #1: Ideal Results Require Ideal Behavior

Insight #2: Purpose and Systems Drive Behavior

Insight #3: Principles Inform Ideal Behavior

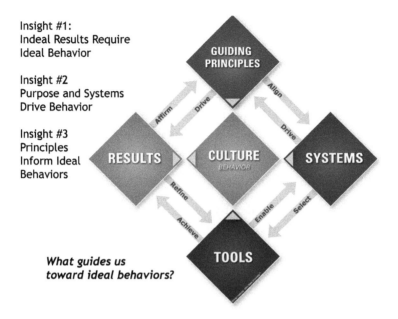

FIGURE 5.2
Three Insights of Enterprise Excellence.

Insight #1: Ideal Results Require Ideal Behavior

As discussed in Chapter 4, one "can" get good results with bad behavior. However, that behavior is not sustainable because people will hate their jobs. Bad behaviors do not create a sustainable culture. However, as discussed in Chapter 2, the enterprise exists to achieve results. Also discussed is the difference between temporary or spot results and sustainable results, which the Shingo Institute considers ideal results.

Previously, the reader learned that the only thing of importance is managing culture. Managing culture requires a shift in the behaviors of all levels of the organization. These behaviors need to shift toward ideal behaviors. So all are on the same page, it is important to define the terms used in this insight:

> **Ideal Results:** *Actions that are aligned, that are both excellent and sustainable, and which demonstrate improvement over time.*
> **Ideal Behavior:** *Actions that create outcomes that produce results and that are both excellent and sustainable.*

The reader should now understand that bad behaviors can give temporary good results, but only ideal behaviors can give sustainable ideal results. To many readers, this may seem obvious, but it still needs to be spelled out and accepted as an insight toward enterprise excellence before moving to the next insight.

The reader learns from Albert Einstein that the definition of insanity is continuing to do the same thing and expecting different results. If one doesn't change their behaviors, they won't change their results. And if the reader wants ideal results, they need to move toward ideal behavior.

Insight #2: Purpose and Systems Drive Behavior

This insight drives toward identifying a greater purpose in life. This principle was previously named, "Beliefs and Systems Drive Behavior;" however, when translated to different languages, this statement was tied to religious beliefs, and that's not what this insight is focused on. This insight drives toward a search for basic ideals. What are those things within that will not change? What is the purpose of one's existence? These are the root causes behind the way one behaves.

This same type of insight, a look at purpose or beliefs, applies to the enterprise. The reader should ask the same kind of questions. What is the purpose of the organization and what drives its existence? If the organization exists only for profit, that would drive one set of behaviors, but if the organization exists in order to improve the life of its employees and its customers, one would see an entirely different set of behaviors. In Chapter 3, the reader learned about behaviors and how these different behaviors manifest themselves. Now the reader will take this understanding to the next level by learning that "purpose" drives the behaviors seen within organizations.

This insight also says that "systems" drive behaviors. The reader can take this to the next level by realizing that even if they have a purpose that focuses on ideal results the supporting systems may motivate behaviors into an entirely different direction. For example, most people live within a budget at home or at work. They watch their spending and the calendar, and they spend accordingly. If they are in a government or academic setting where their budget drives the spending, what happens when money is left over? In these environments, they either spend it or lose it. The current system of budgeting forces them to spend money they should save so they won't be cut short next year when they in fact might actually need the funds. The result is that they stock up on inventory they don't need, just to make sure the entire budget is spent.

What if the opposite were true? What if the system caused one to save when they can? Then when hard times come, they would not have to cut. If the system is correct, it will drive correct behavior and move the needle toward ideal results. Below Dr. Plenert shares couple of examples of organizations who seem to have had correctly focused purposes, but whose systems unintentionally resulted in poor behaviors.

We talk about how systems drive behavior and how important it is to align systems to principles. Have you ever noticed in your personal life or work environment how a policy, or a system focused around a policy, keeps growing? It keeps getting bigger, more complex, and more difficult to administer. We've experienced that numerous times.

One example of systems failure occurred at a factory that had horrible quality problems. Their reject rate was enormous and the CEO asked us to visit the facility and review their quality systems in order to find the solutions. They had quality posters and quotes hanging throughout the facility. They had a quality department.

After taking a tour and listening about how quality conscious they were, we asked them the telling question, "How are your employees measured?" Their response was that they were measured on units produced. So I asked the follow-up question, "Who fixes the quality issues?" And the answer was that the quality department analyzed the failures at the end of the production process and fixed the problems. In this system, the line workers were not responsible for or rewarded for quality. They were rewarded for pumping out units. All that mattered to them was the number of units produced, regardless of whether they were good or bad.

The reward system in this facility drove the undesirable behavior of ignoring quality. Failures were never connected to the source of the problem. The employee reward and measurement system was then changed to measure the quality of the units produced and magically the defect rate shrunk dramatically. Suddenly, employees became more conscious of the quality they produced because it directly affected their paycheck.

Another example of poor systems causing poor results occurred in the purchasing office of one of the world's largest oil and gas conglomerates. In this case, because of a history of inaccurate purchase orders, a series of controls were put in place to make sure that all purchase orders were properly vetted. A purchase order now required 16 approval signatures before it could be released.

The process of routing this document often took six to eight months. Our role as consultants to this organization was to streamline the process, and we started by asking each of the 16 signees what they looked for on the document in order to determine whether or not to sign it. We were surprised when every one of the 16 told us that all they looked for was that another specific individual had signed the document. If so, they were sure the purchase order had to be correct.

In the end, they had 16 control points for every purchase order, all of which created waste, time delay, and none of which added value, since none of them actually checked the document. The failed system drove this behavior because it caused a false sense of security for all the signees.

Were either of these examples following the Shingo Guiding Principles *of creating value for the customer or assured quality at the source? Of course not. It would've been better if only one trusted purchase order signee, who knew they were responsible for the accuracy of the document, reviewed the document carefully rather than having 16 signees who didn't care.*

This reminds us of a Lean philosophy that states, "Always make the right thing easier to do than the wrong thing." A system that is too complex can easily be defeated. It is important to recognize the connections between principles, systems, and behavior. Systems drive behavior. And even with the correct principles, if we don't have the proper supporting systems, we won't get the desired ideal behaviors.

The reader should identify systems that are not correctly focused within their organizations. This is not uncommon. Improperly focused systems can be found in even the best-run enterprises, including Toyota. But returning to enterprises in the West, far too often systems contain metrics as part of the system, and these metrics have been incorporated as part of an accounting or record-keeping process. It is easy to forget that metrics are not for data collection. Employees can make any measurement look good if they think the measure has an effect on their pay or performance rating. It becomes important for an organization to select metrics that motivate the desired response. Remember, metrics are the primary tool that employees use to determine what's important to management and leadership. Poorly designed metrics send the wrong message, so don't be surprised if in the end the wrong results appear.

Another key point to remember about metrics is that too many metrics is as valuable as no metric. When employees are confused with too many metrics, they end up deciding for themselves which metric to focus on, and it may not be the one that management wants them to focus on. So, how important are metrics in the system? They communicate everything. Associates will quickly say, "Tell me how you measure me, and I'll tell you how I perform?" Metrics drive response. For example, if the reader measures how long they are on the phone at a call center, then they tend to rush the customer rather than satisfying them. What is a better metric?

Length of the call, or inquiry resolution (no callbacks) and responses to the post-call surveys? Wouldn't a better objective be, "How does one reduce the number of calls?"

Another example is that simply hanging posters throughout the facility that say things like "Quality is Job 1" or "Safety First" and then measuring employees on throughput results (units produced) thereby encourages associates to ignore the posters and focus on their pay checks. The reader should think back on their own organization. Can the reader identify examples where systems drive behaviors that are not desirable?

Insight #3: Principles Inform Ideal Behavior

In Insight #2, purpose drives behavior. Purpose is founded on principles. Here is a definition:

> *A principle is a foundational rule that has an inevitable consequence.*

With the insight that purpose and principles are interconnected, the Shingo Institute searched out a set of principles that were purpose-focused, which have the characteristic of being:

- **Universal and Timeless:** Principles apply everywhere. They have always existed.
- **Evident:** One can't invent principles but can *discover* them through research and study.
- **Govern Consequences:** Regardless of one's understanding of the principle, they are subject to the consequences of that principle.

Having identified the principles, and understanding that these principles need to be adopted if one is to drive the enterprise toward excellence, the reader needs to carefully consider how each of the principles, understanding that they are ideal principles, drive toward ideal behaviors.

The Shingo Institute has specific principles that will help to develop ideal behaviors. These in turn will help develop enterprise culture. The remainder of this book will help the user understand what each of the principles is and how it guides toward ideal systems, which in turn drives toward ideal behaviors. Principles inform ideal behavior because they help individuals to see the positive and negative consequences of their behaviors and allows them to make more informed decisions.

THE O.C. TANNER STORY OF HOW PRINCIPLES INFLUENCE ENTERPRISE-WIDE BEHAVIORS

Gary Peterson, executive vice president of supply chain and production at O.C. Tanner Company in Salt Lake City, Utah, USA, shares why O.C. Tanner has adopted the *Shingo Guiding Principles* and how these principles have influenced operations.

> *This is our strategy wall. It helps communicate the way we show our people what really matters and what they can do to contribute to the overall well-being of the company. And it all starts with the principles. These are the principles we've gotten from the* Shingo Model *which mean an awful lot to us. We put them up here in the white words and these are the principles that our people really connect with: lead with humility, respect every individual, assure quality at the source, seek perfection, create constancy of purpose. Those kind of ideas. These are bedrock for us. And in the black letters here we've written a description of what we mean by the principle.*
>
> *And I'm convinced that you go on the floor, you talk to any employee here—I don't know that they'll necessarily rattle off the principles—but if you talk to them about respecting individuals, they'll know what you're talking about it, and they'll feel about it the way that we've written about it here. And we try to reinforce this because we find this is foundational to all of the things that happen.*
>
> *So having the principles, then we decide: Well, what is it that we need to have in place in the teams, on the floor, throughout the organization, so we can support these principles? And that's where we come up with the systems. Like the coaching and training system. Not surprisingly, the coaching and training supports the idea of respecting the individual. It teaches people how to embrace scientific thinking. It helps them understand the importance of seeking perfection. It reinforces constancy of purpose. All the different systems help us accomplish the principles. That's why they exist.*

O.C. TANNER
appreciate™

Gary Peterson
Executive Vice President, Supply Chain & Production
O.C. Tanner

RECAP

Enterprise principles, defined by leadership, inform individuals of what behaviors should look like. If the principles are ideal, then the behavior will also be ideal. The *Shingo Model* offers a structure that enterprises all over the world have utilized to drive their organizations toward those ideals.

Reflecting back on the bridge (Figure 5.3) toward enterprise excellence, culture supported by behaviors is central in the development of enterprise excellence. The top of Figure 5.2 shows that guiding principles need to align with systems. Individuals support the guiding principles by exhibiting behaviors that drive ever closer to the ideal. Systems require tools. These tools support and enable the systems. From the tools, through a governance process, results are achieved, and based on the gap in results, the tools are refined. The loop is then closed when the results affirm the guiding principles. In other words, correct behavior occurs when moved in the right direction causing results to become more stable and predictable.

Now at this point, it is time to finish the construction of the conceptual bridge (Figure 5.3). The reader needs to break the enterprise free from the mold of using the tools of organizational change, to become a culture of continuous improvement. One no longer sees change as something that

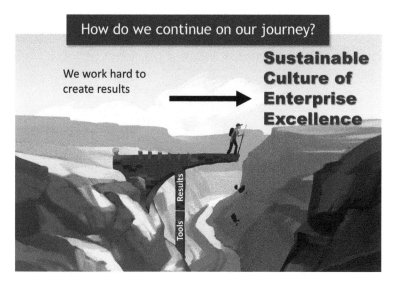

FIGURE 5.3
How Do We Build Sustainability?

occurs once a year, or when faced with a specific problem, or when the opportunity demands something different. In most cases, forced change is change that comes too late. In the Shingo culture, one constantly looks for improvement opportunities. One becomes culturally ingrained in thinking and looking for improvements. The culture becomes so ingrained that one can't even walk into a home or a restaurant without mentally analyzing what they're doing wrong and how it could be done better. Improvement activity is key in one's job. Change is constant, whether one prepares for it or not, and it's better to manage change before it forces unplanned and undesired changes.

Since the *Shingo Guiding Principles* are the foundation that drive results, the reader needs to identify what these guiding principles are, and how they change the way they look at their business, culture, and individual behavior. And that becomes the subject of the chapters to come.

> *The only thing of real importance that leaders do is to create and manage culture. If you do not manage culture, it manages you, and you may not even be aware of the extent to which this is happening.*[*]
>
> **Edgar Schein**
> *Professor, MIT Sloan School of Management*

---

[*] Schein, E. H. *Organizational Culture and Leadership: A Dynamic View.* San Francisco, CA: Jossey-Bass, 1985, 1992.

6

The Shingo Model

Create constancy of purpose toward improvement of product and service, with the aim to become competitive, stay in business, and to provide jobs.*

W. Edwards Deming

THE SHINGO MODEL

The *Shingo Model* has gone through numerous iterations before arriving at the Model in Figure 6.1. And, true to its own principles of continuous improvement, it will probably migrate through even more iterations in the future. But for now, this is the latest and greatest version that the Shingo Institute has to offer.

Looking at the Model, there is integration of all the concepts discussed in this book up to this point. On the left is Results, as discussed in Chapter 2, which is the reason for the existence of any enterprise. At the bottom is Tools, which was identified as the mechanisms used to accomplish the desired results. The arrows between Results and Tools stress that tools help achieve results; and simultaneously as the results that one tries to achieve change, this may in turn cause a refinement of the tools that are used, either selecting a different set of tools, or using an existing set of tools in a different way.

Moving to the right side of the diagram is Systems, and it is systems that drive behavior, which in turn reflects on the culture of an enterprise. Between Tools and Systems are arrows that show that tools are used to enable and support the systems, and that systems, just like results, define and select the tools that are utilized.

---

* Deming, W. E. *Out of the Crisis.* Cambridge, MA: Massachusetts Institute of Technology, 1988.

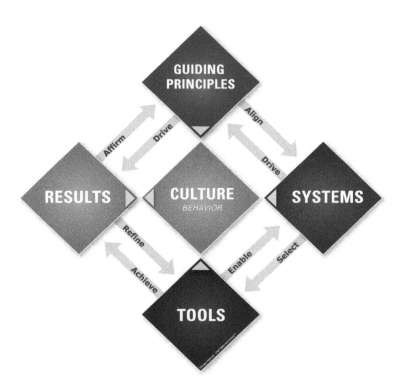

FIGURE 6.1
Shingo Model.

Continuing this journey around the *Shingo Model*, the next box encountered is the Guiding Principles. These have been the primary subject of Chapter 5, and exploring each of them will become the detailed conversation in Chapter 7. Guiding principles, as already discussed, are the substance which supports the purpose and evidence of a culture. Guiding principles drive behavior. Between Systems and Guiding Principles are arrows that graphically show the interconnectedness between these two boxes. Systems drive guiding principles and in turn guiding principles are the foundation used to align systems toward the ideal behaviors one is searching for.

The arrows that connect Results and Guiding Principles show that results affirm a solid foundation of guiding principles and that one needs to adhere to those principles. In turn, the guiding principles, through systems and tools, drive one toward the ideal results that one is attempting to achieve. Guiding principles are the foundation that dictate the consequences or results.

The last box in this diagram, the one in the center which overlays all the others because it is the most important, is Culture. As already discussed numerous times throughout this book, culture is the foundation that drives enterprises toward excellence. Culture, as supported by and manifested in behaviors, finds itself as the center point in the development of enterprise excellence. As Edgar Schein said, "If you don't manage culture, you'll find that culture manages you."

DIMENSIONS

Using the Insights, there are four major categories of principles and these are incorporated into the *Shingo Model*. The categories are Cultural Enablers, Continuous Improvement, Enterprise Alignment, and Results. Breaking these categories down even further are 10 guiding principles.

The *Shingo Guiding Principles* are broken into four logical groups or "Dimensions" following Shigeo Shingo's direction to "think in terms of categorical principles." The bottom, foundational dimension is called Cultural Enablers and it includes two principles. Its focus is on people, which includes all the people within the organization, whether at the leadership level, the managerial level, or the associate level. But it goes beyond looking out at all the stakeholders that are involved with the enterprise, including the value chain that incorporates customers and suppliers, and the community in general.

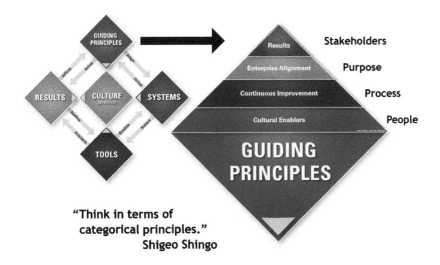

The next dimension moving up the pyramid of *Shingo Guiding Principles* is the Continuous Improvement dimension. This dimension includes five principles and focuses on the processes both within and without the organization. Next, moving up the pyramid is the Enterprise Alignment dimension. This dimension focuses on the purpose of the enterprise. Finding an organization's purpose was discussed in Chapter 5 regarding the Insights. This dimension, with its two principles, strikes at the heart of an aligned, enterprise-wide purpose that everyone within the organization feels comfortable with and in sync.

The last dimension in the *Shingo Model* is the Results dimension. Within this dimension, there is only one principle, and that is a focus on the customer.

THE PRINCIPLES

The 10 *Shingo Guiding Principles* will be detailed in Chapter 7. However, this chapter introduces these 10 principles by each of their dimensions so the reader can get a clear picture of the complete Model. As each principle is introduced, ask if these principles fit the criteria stated earlier. In Chapter 7, the reader will see a detailed explanation of what these principles mean and why the Shingo Institute has chosen to include them in the *Shingo Model*. The criteria that each principle needs to satisfy are that they are:

- **Universal and Timeless:** Principles apply everywhere. They have always existed.
- **Evident:** One can not invent principles but can *discover* them through research and study.
- **Govern Consequences:** Regardless of one's understanding of the principle, they are subject to the consequences of that principle.

With these criteria in mind, the *Shingo Guiding Principles* are as follows:

Dimension 1: Cultural Enablers
- Respect Every Individual
- Lead with Humility

Dimension 2: Continuous Improvement
- Seek Perfection
- Embrace Scientific Thinking
- Focus on Process
- Assure Quality at the Source
- Flow & Pull Value

Dimension 3: Enterprise Alignment
- Think Systemically
- Create Constancy of Purpose

Dimension 4: Results
- Create Value for the Customer

At this point, the reader is ready for Chapter 7, which will explore and explain the meaning behind each of these *Shingo Guiding Principles*.

*A cardinal principle of total quality escapes too many managers: you cannot continuously improve interdependent systems and processes until you progressively perfect interdependent, interpersonal relationships.**

Stephen R. Covey
Author, Leadership Authority & Organizational Expert

---

* Covey, S. R. *7 Habits of Highly Effective People.* New York, NY: Simon & Schuster, Inc., 1989.

7

The Shingo Guiding Principles

Without constant attention, the principles will fade. The principles have to be ingrained, it must be the way one thinks.[*]

Taiichi Ohno
Father of the Toyota Production System

THE SHINGO GUIDING PRINCIPLES

Chapter 6 broke down the elements of the *Shingo Model*. The reader learned about the four dimensions that contain the 10 *Shingo Guiding Principles* (see Figure 7.1). This chapter explores each of these principles in greater depth. Some of the principles are fairly obvious and easy to understand, but others need a more detailed explanation. The reader is encouraged to go through the explanation of each principle and confirm how these principles satisfy the requirements placed on principles in general. Principles are as follows:

- **Universal and Timeless:** Principles apply everywhere. They have always existed.
- **Evident:** One can't invent principles but can *discover* them through research and study.
- **Govern Consequences:** Regardless of one's understanding of the principle, they are subject to the consequences of that principle.

The reader needs to satisfy that these are indeed the 10 principles they want to use as the foundation for their culture. With that confirmation,

[*] Liker, J. K. The Toyota Way. New York, NY: McGraw-Hill, 2004.

FIGURE 7.1
Shingo Dimensions.

the Shingo structure can then help the reader implement a methodology that will move current behaviors closer to ideal behaviors, and thereby give structure to the culture the reader is trying to build. But all of that comes later. First, it is important to understand the principles.

The 10 *Shingo Guiding Principles* came from research by many great Lean thought leaders. The Shingo Institute did not invent them, but they have been proven to boost enterprise excellence and create sustainable results. This chapter introduces and explains the 10 *Shingo Guiding Principles*, broken out in each of their respective dimensions. There are four major dimensions of these principles: Cultural Enablers, Continuous Improvement, Enterprise Alignment, and Results. Breaking these dimensions down even further are the 10 *Shingo Guiding Principles* as shown in Figure 7.1.

CULTURAL ENABLERS DIMENSION

The Cultural Enablers dimension focuses on two principles that are critical in establishing the culture of any enterprise. These are not the only principles necessary in creating a company culture, all the principles are necessary, and all are interdependent, which is to say that none of them stand alone. There is crossover between many of the principles. They work together to achieve the desired culture. However, these two principles in this dimension focus specifically on cultural elements within the organization. Here are some testimonials that successful enterprise excellence companies have made about the Cultural Enablers principles:

Cultural Enablers are made up of two principles of the Shingo Model *and these principles are foundational to the rest of the* Shingo Model *dimensions. Dimension 1 includes lead with humility and respect every individual. And both of those are important and supported by many concepts that help us as leaders, as managers, and as associates, identify what ideal behavior looks like under each of those principles.*

Sheila Montney
AVP—Life, Health and Mutual Funds Product Lines,
State Farm Insurance—Bloomington, Illinois

Leading with humility is probably one of the most difficult principles for leaders to accommodate because leaders have a propensity to come into an organization, solve problems quickly, and really have all the answers when it comes to going into the workplace.

And leading with humility requires this vulnerability for leaders to really go into the organization, support members, and drive member engagement.

Tony Hayes
Director of Continuous Improvement, Haworth
Grand Rapids, Michigan

I think performance starts with the leaders of the organization. They set the tone of the organization. Their ethical guidance, their display of values of the organization are what drive the organization and set the tone for the culture of the organization. And if they walk the talk of the values of the organization, it's easy to build the trust and build a high-performing culture. If they misstep, culture disappears overnight.

Harry Hertz
Director Emeritus, Performance Excellence, Baldridge
Gaithersburg, Maryland

I was brought up in Africa and through family connections, there used to be a greeting in Zulu called "Sono Bono," which translated means "I see you." But what it actually means is, "I see ALL of you." I see your past, I see your current, but I also see your potential. For me, that is why I really align to the principle of respect every individual. Because it's about not only seeing the individual as they are today, but their potential, and seeing how much capability and potential they have for the future.

The importance for that principle is how it links and drives the right behaviors of the individual and therefore, what that impact is on the culture. Because the culture is what drives the organization, and ultimately, the results.

Morgan Jones
Head of Group Productivity, Commonwealth Bank of Australia (CBA)
Sydney, New South Wales, Australia

My core message in trust is so tied into the Shingo Guiding Principles *of leading with humility and respect for every individual, because the foundation for this is integrity. I define integrity as not only just having honesty, but also humility and courage, because we have to be humble and open to principles. There are principles that govern, and being open to that is important. Any great leader that starts with humility is in a different place than a leader that leads out with arrogance or hubris.*

Respect for individuals, one of the guiding principles of the Shingo approach, is also so vital. It's one of the key behaviors that builds trust: demonstrating respect. It's how you treat everybody. Now the opposite is when you disrespect people. The dangerous counter effect is when we show respect to some, and not to others. We tend to show respect to those who maybe can do something for us, but don't show that same respect to those who maybe can't.

But how we treat the one has an effect on the many. A doctor who disrespects a family member could lose the trust of a patient. But a doctor who shows great respect of a family member could earn the trust of a patient. How you treat the one has an effect on the many, and that's never more apparent than in respect—respect for every individual.

This guiding principle of the Shingo approach is also foundational to my approach on building trust. Those two cultural drivers, behaviors of leading with humility and respecting every individual, are really the first two elements of my model of integrity and intent. Integrity encompassing the humility, and intent encompassing respect.

With those two pieces in place, then we're able to add to it other pieces, and they're viewed through a different context than if we don't have those two pieces in place. So, I think there's a great alignment between trust and humility and respect.

Stephen M. R. Covey
Co-founder, Speed of Trust Practice, Utah

The *Shingo Model*'s Cultural Enablers dimension includes the following principles:

Respect Every Individual

Respect for every individual is manifested when organizations structure themselves to value each individual as a person and nourish their potential. How one treats another person is seen by everyone else and affects how they feel now and how they anticipate being treated in the future. Disrespecting just one individual sends a wave through the entire organization causing all employees to wonder if they may become the next target of disrespect. And the culture of the company is seriously damaged.

When people feel respected, they give not only their hands but also their minds and hearts. This includes individuals who are often not an active part of operations but should be. Individuals from support functions such as safety systems, health and welfare systems, continuous improvement systems, and so on need to be engaged just as much as production line workers. All need to be respected.

Some extreme examples of the ultimate bad guys would be Hitler, Stalin, Pol Pot, and a dozen more. Did these individuals show respect for every individual? Of course not, but what is more important is to look at the consequences of their behaviors. So how does the caste system or apartheid fit into this model? And of course, these individuals or these racially based systems do not fit in. These systems do not show respect for the individual. They do just the opposite. They demonstrate disrespect.

People want to be respected. They want to be engaged. They want to have meaning in their life. This principle is universal and timeless. It has always been that way. It was true 2,000 years ago, and it will be true 2,000 years from now. From an enterprise level, if employees are engaged, the company does better.

Lead with Humility

Humility manifests itself when leaders seek out and value the ideas of others. They need to be willing to change when they learn something new. They should encourage employees to submit ideas and respond to suggestions positively. Effective leaders recognize that the people doing the work are the ones that understand the work best, and are the best at identifying opportunities for improvement. They must trust others to make good decisions, and recognize there are negative consequences for not paying attention to input. This principle does not mean that leaders are not bold in implementing programs aligned with the strategic objectives of the organization. Rather it means that leaders are not islands and they recognize that everyone in the enterprise has creative insights and ideas.

Organizational and personal growth is enabled when leaders work to bring out the best in those they lead. All growth requires vulnerability (the willingness to allow others to change the organization, giving them the authority to have power over things that influence a leader's performance). Individuals acknowledge that they don't know everything and are willing to learn. With this model in practice, decision-making moves to lower levels of the organization.

To better understand this principle, it is important to draw out the associates, identify how they interpret the behavior of their management, and ask them questions like, "When you have an idea, does management act on it?" That doesn't necessarily mean they do what was suggested, but there needs to be some kind of response that management actually acknowledged the idea.

What are the best ways within an organization to lead with humility? To start with, management needs to recognize that they may not be an expert at everything. Even if they at one time ran the production equipment themselves on the floor, things change (the machinery, products, temperament of the equipment, etc.). Many often wonder, "Why does management not have a problem admitting that they don't know everything about IT and are willing to call in an IT expert, but when it comes to production they feel the need to be the expert?" In most cultures, the employees have learned to expect leadership and management to be the expert, and like sheep they follow whatever they are told to do without question. Humility is the ability to admit that people closest to the process actually know more about a product or service than the leader or manager does. Leaders need to be open to hear and consider what others say.

Questions the organization needs to ask include, "Do people want to be respected? Do they want to be engaged? Do they want to have meaning in their life?" The answer to these questions is of course, "Yes!" Empowering them to make decisions and to have their decisions respected moves everyone in this direction.

Some organizations challenge this principle by asking, "Can one be a good leader and still be humble? Doesn't a good leader need to be demanding?" They often cite the example of Steve Jobs as someone who was demanding and hard to work for but still accomplished results. Bad behavior can still achieve short-term performance results. But is that a type of culture that is sustainable in the long term? This brings up the question, "Would one rather work for a person with a strong vision who is not humble, or a person who has no vision but is humble." The answer, of course, is that one wants a principled leader that is both demanding and at the same time humble. That's the definition of a good leader. When a leader is not principle based, a "blame culture" is created.

CONTINUOUS IMPROVEMENT DIMENSION

The Continuous Improvement dimension dives deeper into the process elements of an organization, such as quality, inventory, flow, problem solving, etc. Under this dimension, there are five principles, and the principles need to include discussions of continuous breakthrough, innovation, and continuous learning.

There are some foundational tools to incorporate into every enterprise before an organization can successfully move forward with continuous improvement. The first of these is 5S. The reason this is so important is because it is impossible to improve an operation or an organization that is disorganized. One cannot standardize something they cannot find.

The second foundational tool to incorporate is standard work. If everyone does the same function differently, then how can the organization improve it? There first needs to be a standard methodology before one can look at options for improvement. Then, once these two fundamental tools are in place, the reader is now ready to progress toward enterprise excellence.

Here are some testimonials that successful enterprise excellence companies have made about the Continuous Improvement principles:

Continuous improvement is always something that intrigues me and the reason why is that we are in the relentless pursuit of improvement. Improvement is intuitive to all of us, specifically, when I think of continuous improvement. Before I thought of it as an end game, so you would improve and be done. But what I've learned from Shingo and what we've focused on as a company is that once we've improved and gotten to that benchmark, we've changed that benchmark.

Troy Blanchard
VP of Production, Cannon Safe, Industrial Pacifico
Tijuana, BC, Mexico

We can't create value if we don't understand it. We can't create value unless we have a baseline. So we start to make changes, we can measure those improvements, I think it's very important to start with measures that matter and then align systems, tools, people, processes so you can achieve those outcomes.

And so it loops back a little bit like 'plan–do–check–act.' You've got to plan for value for customers. You've got to do that work. You've got to check by re-measuring—did it give you the right outcome?—and then adjust or act as necessary. And then do the loop again.

Michelle Lue-Reid
GM Group Productivity, Commonwealth Bank of Australia (CBA)
Sydney, New South Wales, Australia

One of the products we have is individual life claims. The value stream that we looked at to improve was the beneficiary calling up on the claim, notification of a death, and then the payment of that claim to the beneficiary. We looked at that process and our lead time on that process was about 45 days from the point of notification to the point of payment.

The goal was to reduce that down. If you think about the beneficiary when they call, a lot of things are going on. A lot of payments are being requested, a lot of expenses in that process, and a lot of people buy life insurance to cover those expenses. So, having a 45-day period from notification of payment could actually cause some hardship to the beneficiary.

Our goal was to look at that process and reduce that time significantly. Our goal was 70%, 80% when we originally went into the transformation process.

As we started thinking about how we seek perfection in a process like that, and thinking about value from the beneficiary's standpoint, the team was able to reduce that from 45 days from notification of payment down to the same day. So the goal of the team today is on the phone call from the beneficiary to get enough information and search for information that will supply what is needed to actually pay-out that life insurance claim to the beneficiary within one call.

Dan Christel
IT Manager, Continuous Improvement, Mutual of Omaha
Insurance Company—Omaha, Nebraska

The *Shingo Model*'s Continuous Improvement dimension includes the following principles:

Seek Perfection

Seek in this case means "pursue." Seeking sometimes has an end point whereas pursuing suggests a process of continuous improvement. Seeking perfection is looking at something beautiful and wanting to be like that. It's often referred to as searching for the True North. There is a need to see what one wants, the goal, and how to work toward it. Seek perfection has its focus on pursuing excellence.

Perfection is an aspiration not likely to be achieved, but the pursuit of which creates a mindset and culture of continuous improvement. The realization of what is possible is only limited by the paradigms through which one sees and understands the world. It's always about getting better and better, never satisfied with good enough, and always looking for opportunities to take leaps forward. This cannot become a baseball bat used for beating people. Rather it's about positive motivation. It's about setting ever higher, yet attainable, goals and then achieving them.

One TPS tool that would help facilitate this principle is incorporating the concept of the eight wastes (the traditional seven wastes of Lean plus an eighth waste, which is a "waste of talent"). Another tool is the use of a suggestion system. Also consider innovation tools like breakthrough thinking and concept management.

Embrace Scientific Thinking

The first step of scientific thinking is to define standard work for the process being analyzed. If one does not have a standard way of doing a job, commonly referred to as standard work, then it becomes impossible to change that standard. Once the job is consistently done the same way, one can then take a closer look at failures or waste in that process. One searches for the root cause of the failures, finds the problem in the system, and fixes the system so the problem doesn't reoccur. In scientific thinking, everyone in the enterprise looks at the problems and generates repeated cycles of experimentation and improvement. Innovation and improvement are the consequences of repeated cycles of experimentation, direct observation, and learning. A relentless and systematic exploration of new ideas, including failures, enables one to constantly refine standard processes.

Tools that facilitate embrace scientific thinking (or embrace the scientific method) include first and foremost standard work with respect to quality, flow, value-added work, etc. It includes Lean, breakthrough (*kaikaku*) improvements, or small step improvement (*kaizen*). Other applicable tools are SDCA or PDCA (standardize, do, check, act or plan, do, check, act) and value stream mapping (VSM).

Focus on Process

All outcomes are consequences of a process. It is nearly impossible for even good people to consistently produce ideal results with a poor process both inside and outside the organization. There is a natural tendency to blame the people involved when something goes wrong or when something is less than ideal, when in reality the vast majority of the time the issue is rooted in an imperfect process, not in a failure of the people. Often it is fear that causes problems to be focused on a person. Who is really responsible for the system that might be failing? The tendency is to blame an associate using the system, especially if management is the one responsible for the system, when in reality it is the process that causes people to fail. It is leadership and management's responsibility to follow the Lean concept of, "Making sure the systems are designed so that it is easier to do the right thing rather than the wrong thing." One needs to design systems that have considered the possibility of failure and are designed in such a way so as to make failure impossible.

Tools that facilitate a focus on process include Lean, Six Sigma, or root-cause tools such as fishbone charts and the "5 Whys."

Assure Quality at the Source

Perfect quality can be achieved only when every element of work is done right the first time. If an error should occur, it must be detected and corrected at the point and time of its creation. Do it right the first time. This goes back to when the processes are designed and developed. Don't pass poor quality forward. Stop the line if there is a quality failure. This principle also includes error-proofing, which goes back to the Lean concept that says, "Always make the right thing easier to do than the wrong thing." Design systems to minimize the opportunity to make mistakes. Another aspect of this principle is the gemba concept, which drives one to "go and observe" the actual work at the location where the work is taking place. This is required in order to understand how the work is performed and to assure that quality is incorporated at the source of the work.

Flow & Pull Value

This is at the heart of TPS, incorporating the entire supply chain. The focus is on creating value and includes customers, marketing, production and operations, logistics, vendors, etc. Toyota measured the time from when the order was placed until the money was in their pocket. They consider the reduction of that time as the measure of increased value. Another measure is total cash-to-cash cycle time as the ultimate definition of success and value. The reader should study the entire value stream and eliminate the waste in all the steps within the process.

Value for customers is maximized when it is created in response to real demand and a continuous and uninterrupted flow. Although one-piece flow is the ideal, often demand is distorted between and within organizations. Waste is anything that disrupts the continuous flow of value.

Dr. Shingo's book on nonstock production stresses one-piece flow and is referred to as the Shingo Production System.* This book is valuable if the reader is interested in a deeper understanding of this concept.

Tools that facilitate flow and pull include standard work, VSM, SMED, poka-yoke, kanban, and 5S.

ENTERPRISE ALIGNMENT DIMENSION

Understanding enterprise alignment means having an unwavering clarity of why the organization exists, where it is going, and how it will get there. This enables people to align their actions, as well as to innovate, adapt, and take risks with greater confidence. This is the true north, vision, mission, and values of the enterprise.

Here are some testimonials that successful enterprise excellence companies have made about the Enterprise Alignment principles:

> *Alignment is when all handoffs in an organization occur properly and everybody sees the next person along the way in producing a good or service as their customer, so that you're in sync with each other and handoffs occur so that the organization operates as a system. We see integration as the next step as it's really going from alignment to integration so there is concurrent planning, concurrent operations, and the whole is more than the sum of the parts.*

Harry Hertz
Director Emeritus, Performance Excellence, Baldridge
Gaithersburg, Maryland

> *It touches every part of the enterprise, your manufacturing operations, your office, your administrative-type functions, HR, IT, and finance. And really it ties everything together as it relates to constancy of purpose, alignment of objectives from top to bottom. And if your organization is aligned in this manner, then you're driving these principles in your enterprise and you'll see this top to bottom.*

Tony Hayes
Director of Continuous Improvement, Haworth
Grand Rapids, Michigan

---

* Shingo, S. *Non-Stock Production: The Shingo System of Continuous Improvement.* Portland, OR: Productivity Press, 1988.

The *Shingo Model*'s Enterprise Alignment dimension includes the following principles:

Think Systemically

Simply put, what happens here either improves or messes up what happens in other parts of the process. Thinking systemically means looking at the big picture. Through understanding the relationships and interconnectedness within and between systems, one is able to make better discussion and improvements. Everything is connected, and the better one understands this connection, the better they are able to make changes that improve the system.

Create Constancy of Purpose

This principle focuses on creating unity. An unwavering clarity of why the organization exists, where it is going, and how it will get there enables people to align their actions, as well as to innovate, adapt, and take risks with greater confidence. This principle requires a methodology for communicating company information to every employee, including goals and metrics.

RESULTS DIMENSION

The focus of most leadership is on results, commonly called KPIs. The principle in this dimension stresses that value needs to be considered from the perspective of the customer, rather than from that of enterprise leadership. Unfortunately, leading indicator KPIs are usually what one uses as measurements of behavior.

It is important to differentiate between leading and lagging indicators. Leading indicators are generally described or are closest to behavior, whereas lagging indicators measure performance results. Therefore, the indicators used to influence and change behavior are generally different from, but connected to, the ones that report the results or KPIs on which leadership is measured. Please do not misunderstand. An organization must get results to succeed. Simply put, how one achieves the results is just as important as the result itself when it comes to sustainability and enterprise excellence.

Here are some testimonials that successful enterprise excellence companies have made about Results:

> *The results come. Instead of always focusing on dollars and savings, it's about focus on the people, external people—our customers—and internal staff. That drives the culture, that drives the results.*

> **Morgan Jones**
> *Head of Group Productivity, Commonwealth Bank of Australia (CBA)*
> *Sydney, New South Wales, Australia*

The *Shingo Model*'s Results dimension includes the following principle:

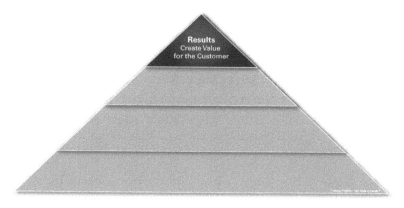

Create Value for the Customer

The focus of this principle is on creating value for all stakeholders. Ultimately, value must be defined through the lens of what a customer wants and is willing to pay. Organizations that fail to deliver both effectively and efficiently on this most fundamental outcome cannot be sustained over the long run. If the customer does not feel value, the organization will lose the customer, and nothing else really matters.

REVIEW

At this point, the reader should have a clear understanding of each of the 10 *Shingo Guiding Principles* and how they are applied. The reader should ask the following questions about each of these principles. Are they:

- Universal and Timeless?
- Evident?
- Govern Consequences?

With this understanding, the reader is now ready to decide if these are the principles they want to incorporate into the cultures of the enterprise. Does the reader want to transform the company's culture to follow these 10 principles? And how does one accomplish this transformation? The next section of this book delves deeper into the how and why of achieving enterprise excellence. It's now time to press forward.

*It was as if we were engaged in car manufacturing in a virtual world and became insensitive to vehicle failings and defects in the market.**

Akio Toyoda
CEO, Toyota
Grandson of Toyota's Founder

---

* Taylor, A. How Toyota lost its way. *Fortune Global 500.* July 12, 2010. Available at http://archive.fortune.com/2010/07/12/news/international/toyota_recall_crisis_full_version.fortune/index.htm

Section III

Knowing Why

8

Thought Leadership

Principles are foundational rules and help us to see both the positive and negative consequence of our behaviors. This fact enables us to make more informed decisions, specifically, about how we choose to behave.[*]

Shingo Institute

ASSESSING CULTURE

Key behavioral indicators (KBIs) provide observers with initial predictors of an organization's culture. KBIs might include: excitement as seen in people's eyes, people moving with a sense of urgency, people making eye contact, people excited to talk about what they do, people approaching leaders to greet them, leaders greeting people by name, leaders engaging with the people they greet, etc. Observers can quickly begin to understand an organization's culture as evidenced by the behaviors they see.

To illustrate this with an example, if the reader's current enterprise culture is focused on firefighting, the culture that exists is one where firefighters become the heroes of the organization. Firefighters are the ones that get promoted. In an enterprise where firefighting is the norm, employees look for and get excited about fires. Some employees even become arsonists so they can be the first on the scene of the fire. However, what would be closer to ideal behavior? Wouldn't it be desirable to not have any fires? Shouldn't one reward and promote people who prevent fires from occurring by improving the situation in such a way that fires can never start? This shift in behavior can often be very traumatic for an organization

---

[*] Shingo Institute, *Three Insights for Enterprise Excellence*. See http://shingo.org/model.

because it begins to shift culture. Especially if the behavior is frequent, intense, exists for an extended period of time, permeates all roles in the organization, and spreads through the entire enterprise and its scope. In this way, behaviors shift culture. This requires a change in the systems that have promoted this culture. It requires a movement away from the traditional thought processes. It requires a principle-based organization with a deep and continued growth in its understanding of the principles. In most cases, this goes far beyond merely the business implication.

To consider a few examples of how a principle-based change in culture affects behavior, the first example takes a look at a performance evaluations system. Everyone has an annual performance review of his or her work during the last year. But is this ideal? No! An ideal performance evaluation would include timely reviews as the work is performed, so corrections can be made rather than waiting until the end of the year and getting hit with a bombshell when it is too late to do anything about it. And it should be more than just a one-directional evaluation. Interactive discussions help employees understand the evaluation and include a better understanding of what improvements are necessary. Additionally, it would be ideal to revisit and update goals as job situations shift. Goals should be aligned with the purpose of the organization. They should be measurable. The example of misused performance evaluation systems shows how systems drive behaviors.

Where does the reader go from here? What does all of this mean? As the reader can see from the Model, the foundation of enterprise culture is at the heart of the entire *Shingo Model*. All the principles need to be supported by the culture. They accomplish this by informing ideal behaviors, or what becomes the behavioral goals. Cultural shift requires a shift in behaviors. Behaviors are driven by systems. Systems need to align with the principle through the ideal behaviors they inform, shifting the culture ever closer to the desired status. In the end, one will most likely need to adjust old systems, create new systems, and eliminate systems that no longer support the desired culture. The reader should want to create a cultural foundation to see customer-oriented results at the top, and enterprise alignment at the bottom. In the middle are all the continuous improvement behaviors, systems, and tools that drive toward ideal results.

> We tend to show respect to those who maybe can do something for us, but don't show the same respect to those who maybe can't. But how we treat the one has an effect on the many.

Stephen M. R. Covey

Looking back at the bridge, what has changed? The left half of the bridge is focused on tools and results. The type of results is KPIs. These are primarily financial in nature and are not principle-based. Focusing only on KPIs means that the enterprise is short-term focused. It's not sustainable. It tends to focus on fighting fires in order to provide short-term results. However, if one completes the bridge by creating a principle-based culture, they focus on KBIs. Principle-based cultures focus on the long term. They shift the way an organization thinks. And when an organization thinks long term, it drives improved KPIs through improved KBIs.

An enterprise that does not operate on principles operates on an ever-growing set of policies. The enterprise becomes overwhelmed with an overabundance of unnecessary control systems that police employees rather than empower them. In the end, the principle-based cultural shift encompasses and consumes all systems, tools, and results, whereas a policy-based structure "controls" systems. That's simply not acceptable because control costs too much, as seen in Chapter 5 where a purchase order required 16 signatures and six to eight months to process. The reader needs a long-term, principle-based culture in order to drive enterprise excellence. Section IV discusses how to transform the enterprise's culture.

How does one implement and integrate principles and use them to change behaviors and systems? First, define what systems and tools are in an organization.

System: A system is a collection of tools or tasks that are highly integrated to accomplish an outcome. It is important for those involved in a given system to not only be aware of their responsibilities and tasks, but also how to go about conducting them. It is not uncommon for organizations to have different isolated systems within the whole. However, making employees familiar with how the various systems work and where they fit in increases familiarity and productivity within the organization.

Tool: A tool can be defined as a single device or item that accomplishes a specific task. It is how the work gets done, not simply through physical means but the methodology by which it is accomplished.

Who builds these systems and tools? The management? No. The people. And they are directed by the company culture. Obviously then one must focus on the cultural behavior because it helps one understand the values and beliefs. Culture is one of the most important aspects of the organization. It is one thing to visit, or observe outside organizations and view how they implement certain systems or use tools to accomplish goals. It is an entirely separate and more important task to understand "why" these systems are in place. It is better to understand the methodology; the reason the culture within the organization has developed the way it has; and how they have a core set of principles, beliefs, or ethos, the employees and managers of a given organization can hold on to and emulate. Cultural Enablers are the foundation of the *Shingo Model*.

At this point, it would be valuable to engage in an exercise that is meant to cement some of the principles down. Please make a four-column spreadsheet where all the 10 principles are listed in the first column. In the second column, predict the consequences to the organization if they follow this principle. In the third column, predict the consequences to the organization if they choose not to follow this principle. Then, in the fourth column, rank the enterprise on a scale of 1–10 on its performance of that principle where a 1 is nonconformance and a 10 is complete conformance. Having completed the fourth column, add that column up and come up with a total. That total represents the percentage of organizational readiness toward becoming a Shingo award-winning company. It's difficult to be completely honest here because the reader may be prejudiced by what they know about the organization. If possible, have several individuals in the organization fill out that spreadsheet (or at least the fourth column, using the first three columns that were created) and compare those results with the reader's results. The differences will be surprising. Chapter 11 will discuss the benefits of an external assessment, but for now the reader will have created their own estimate of the organization's performance.

To conclude this section, here is an example of a Shingo Prize-winning company and their successful utilization of the *Shingo Guiding Principles*.

> *Here at Autoliv, understand that we are in the perfection business. We get paid for saving lives. We make life-saving devices every day that are the difference between someone being able to walk away from the accident, or maybe not. And so when we look at what drives improvement, we need to look at sustainable improvement, because we're talking about perfect performance every day.*
>
> *And every day we challenge ourselves to be better than we were the day before. How do we do that? You do that by reaching deep into people. And by linking with values that drive their behavior and the base for those kinds of values or principles, we try to grab one of those principles and we embody it in our model that "we save lives." Every day we come to work and we save lives.*
>
> *We're in a very tough business. In the automotive business, if you don't improve, you don't exist. With every person that comes into that door, we have multidimensional capabilities. And our job as management is to create and sustain the environment that fosters those abilities to be expressed. Expressed as improvement ideas. Expressed as them developing their workplace so it is owned by them.*
>
> *Now, in a role as one of our top managers, my job is that we keep focusing on those things that will sustain our ability to add value over time in a better way. In business schools, you might be told, you have to drive lower costs by lower overhead, lower labor costs, or lower material costs. Well, all those are great but you don't do those directly. You do those through investing and raising the skills and abilities of your people. And if you do that and do it well, then they do all those things and they do it better than you know how to do because it's their house. It's their job. And they do it very well.*
>
> *So, as management at Autoliv, our job is to hold up those principles that sustain their behavior on a daily basis. And as we do that, as we are seeking perfection through employing their ideas every day in the workplace, by showing that we trust them by implementing their ideas, we are then in a position to accomplish the goals of the business: save more lives every day by implementing the ideas of our people. And by doing that, we all win.*
>
> *You can operate in the short-term by just pursuing results, and you can get results in the short-term. But it's a sugar high. It doesn't last. It's not sustainable. And so what it degenerates into is a program. Just another program. But the way we do it, through investing in our people and sustaining the culture that we have, by doing that, we invest in our future, we*

invest in the future of our people. And we grasp that future and we make it our own.

Basing improvement on principles is a long-term perspective. It is looking beyond the horizon. Basing improvement on anything other than principles is, by its very nature, finite. It cannot sustain itself in the long-term.

I measure success by the light in their eyes. When you see the spirit through the eyes and that spirit is dynamic and vibrant and on fire, what else do you need to see?

Thomas Hartman
Former Director, Operations, Autoliv
Brigham City, Utah

KNOWING WHY

A two-year-old asks the question "Why?" on average 160 times a day. People lose that as they get older. Why? Have people lost the ability to be curious? Or are they just suppressing it? School teaches one to sit in the chair and shut up. People lose the desire and confidence to ask "Why?" In far too many of the enterprise cultures of the world, this is reinforced because leadership and management reinforce an attitude of "Do what you're told and don't ask questions." If the reader is to follow the *Shingo Guiding Principles*, they need to regain the focus and confidence for all employees throughout the enterprise to ask, "Why?" For example, Walmart started with a focus on "why" which was to improve customer service and employee satisfaction. Over time, it shifted closer to the "what" causing less of a focus on both employees and customers. The "what" drove it to lower prices. A disconnect occurred and as a result they now have increased employee dissatisfaction and they have communities that boycott the construction of a Walmart in their area.

What is the difference between the "why" and the "what?" The difference is in participation and involvement. "Why" suggests humility, curiosity, and an interest in being better. "What" suggests an edict or a decree from the top down. So the reader needs to ask, "What difference does it make when people know why?" The difference is that employees want to be a part of the big picture. They want to participate in the process, not just be an easily replicable cog in the wheel.

PRINCIPLES INFORM IDEAL BEHAVIORS

Referring to Figure 8.1 is an example of where employees at the different levels of the organization should apply their efforts. These numbers are fairly arbitrary and are not to be considered precise measures because different enterprises will apply these principles in different ways, but as a guideline, the Shingo Institute proposes the following distribution of effort:

Guiding Principles—Leaders spend 80% of their time.
Results and Systems—Managers spend 80% of their time.
Tools—Associates spend 80% of their time.

With these approximations one needs to ask the question, "Does ideal behavior vary by role?" The answer of course is "no" since ideal behaviors are universal. Then one needs to refer back to the third Shingo Insight

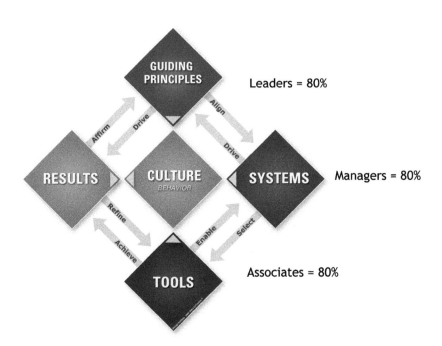

FIGURE 8.1
Critical Areas of Focus.

that was discussed in detail in Chapter 5 and ask, "How do 'Principles Inform Ideal Behaviors?'" It starts at the top with leaders modeling ideal behaviors. Until the organization's employees are able to witness a commitment to ideal behaviors at the leadership level, they won't take the shift to *Shingo Guiding Principles* seriously. As part of this process, leadership needs to make sure there aren't any legacy systems that conflict or contradict the principle-focused systems one is trying to exemplify. Eliminate systems that conflict and replace them with systems that support ideal behaviors.

As one moves toward ideal behaviors they need to recognize and avoid the default excuse of "operator error." Almost always it is the system (or lack thereof) which causes the operator error. The system should be changed in such a way as to prevent the error. In a well-designed system, the operator would not have been able to make the error. Therefore, it is the system that requires the reader's focus, not the operator.

As an exercise, try to go through the following tables of principles and see how each of these principles manifest themselves at each of the three levels of employees within the enterprise. Ask if principles do indeed inform ideal behaviors. The following list offers up examples for five of the principles and the reader should carefully consider each of these and look for their own examples that exemplify each of these principles. Additionally, these tables ties systems and tools with each of the principles:

Principle:	Respect Every Individual
Leaders:	Leaders always acknowledge the specific behaviors they see that are close to ideal.
Managers:	Every manager makes certain that large group meetings begin with a safety briefing.
Associates:	Associates demonstrate an eagerness to learn new skills, take initiative, and share their learning and success with others.
Systems:	Recognition Safety Communication
Tools:	Recognition events and appreciation cards Proper safety equipment Newsletter

Principle:	Lead with Humility
Leaders:	Leaders listen and respond and are not dictatorial.
Managers:	Managers encourage good behaviors through an appropriate reward system.
Associates:	Associates take more ownership to resolve issues.
Systems:	Feedback and recognition Problem-solving
Tools:	Suggestion boxes Communications tools

Principle:	Embrace Scientific Thinking
Leaders:	Leaders set the tone by using the same problem-solving tools as the rest of the organization.
Managers:	Managers teach problem-solving tools to the people and approve a space for associates to use the tools.
Associates:	Associates are flexible, willing, and have a desire to ask why. They participate in continuous improvement.
Systems:	Communication Problem-solving
Tools:	PDCA Newsletter Reports Cross-functional teams Tracking progress

Principle:	Flow & Pull Value
Leaders:	Leaders concentrate on one thing at a time.
Managers:	Managers communicate customer requirements.
Associates:	Associates are eager to share information between steps in the flow.
Systems:	Improvement management Material planning
Tools:	KPIs Employee recognition

Principle:	Thinking Systemically
Leaders:	Leaders encourage cross-functional participation (recognize individuals).
Managers:	Managers promote collaboration.
Associates:	Associates are open-minded and willing to change (participation).
Systems:	Problem-solving Improvement management
Tools:	"Go and observe" walks Public presentations

Five principles are left for the reader to create their own table. These principles are:

Seek Perfection
Focus on the Process
Assure Quality at the Source
Create Constancy of Purpose
Create Value for the Customer

Once the reader has reviewed the five principles listed as examples above, and has also created tables of their own for the five remaining principles, they should have a solid understanding of the third Insight and see how "Principles Inform Ideal Behaviors."

> *Personal mastery is a discipline of continually clarifying and deepening our personal vision, of focusing our energies, of developing patience, and of seeing reality objectively.*[*]

Peter M. Senge
Author

---

[*] Senge, P. M. *The Fifth Discipline: The Art & Practice of the Learning Organization.* 2nd edn. New York, NY: Random House, Inc., 2006.

9

New United Motor Manufacturing, Inc.: A Case Study That Informs the Shingo Model

As discussed in previous chapters, the Shingo Institute defines culture as the accumulation of behaviors within an organization. In order to better understand how principles can be observed in the behaviors of people in an organization, it helps to "observe" behaviors in action through a case study. A case that has informed the development of the *Shingo Model* is New United Motor Manufacturing, Inc. (NUMMI), which was the first major investment by Toyota outside of Japan.

THE HISTORY

General Motors opened an assembly plant in Fremont, California in 1963. Within 15 years, the plant employed over 7,200 workers, but by 1982 the plant was closed. Charles A. O'Reilly, III and Jeffrey Pfeffer, authors of *Hidden Value: How Great Companies Achieve Extraordinary Results with Ordinary People,* said:

> *The reasons for closing the plant were sound—GM-Fremont ranked at the bottom of GM's plants in productivity and was producing one of the worst quality automobiles in the entire GM system—not an easy task in the early 1980s. A militant union averaged 5,000 to 7,000 grievances per labor contract. The plant was characterized by high sick-leave counts, slowdowns, wildcat strikes, and even sabotage. Daily absenteeism was almost 20% ...*

The result was a climate of fear and mistrust between management and the union.*

On the basis of the exercise in Chapter 3 that described poor behaviors commonly present in cultures such as firefighting, fear, power hungry, survival, etc., the behaviors presented in this case so far were very predictable given the plant's culture. It is not unusual that one would want to change the behaviors evident at the GM facility described above. What is unusual is the confidence and ability to do so. Toyota must have seen an opportunity in this GM facility, because the two organizations pursued a joint venture together in 1984 called NUMMI.

Under the terms of the joint venture agreement, Toyota and GM agreed to invest roughly $100 million each with another $200 million in debt for the new organization. NUMMI agreed to recognize the union, recall the workforce that was laid off when GM closed the plant, and pay the employees union-scale wages. The United Auto Workers (UAW) union agreed to embrace TPS to increase the flexibility of work rules and simplify the numerous job classifications, and Toyota agreed to utilize the same 25-person union bargaining committee that existed under the old GM system.† Toyota then assumed responsibility for all plant operations, including product design and engineering as well as marketing, sales, and service for vehicles with the Toyota marque, and GM assumed responsibility for the marketing, sales, and service of GM-badged vehicles.‡

The Workforce

Prior to closing the plant, the GM-Fremont workforce was considered the worst in the U.S. automobile industry. Bruce Lee, who ran the UAW western region and oversaw the Fremont plant, said, "It was a reputation that was well earned. Everything was a fight. They had strikes all the time. It was just chaos constantly."§

Rick Madrid who built Chevy trucks at the plant said there was a lot of booze on the line, but as long you did your job management really didn't care. He would bring a thermos of screwdrivers and drink while he

* O'Reilly III, C. A. and Pfeffer, J. *Hidden Value: How Great Companies Achieve Extraordinary Results with Ordinary People.* Boston, MA: Harvard Business School Press, 2000, p. 182.

† O'Reilly and Pfeffer, p. 183.

‡ O'Reilly and Pfeffer, p. 184.

§ Langfitt, F. The end of the line for GM-Toyota joint venture. March 26, 2010. Available at www.npr.org/templates/story/story.php?storyId=125229157.

mounted tires. "It wasn't just drinking and drugs," Madrid said. "People would have sex at the plant, too."* Many workers disliked management so much that they would sabotage vehicles by putting Coke bottles inside the door panels to rattle and annoy customers. Some would question how people were able to keep their jobs, but under the union contract workers practically had to commit fraud to get fired.

Absenteeism was a serious problem. Billy Haggerty, who worked in hood and fender assembly, said some mornings there wouldn't be enough workers to show up to start the line. Managers would "go right across the street to the bar, grab people out of there and bring them in,"† Haggerty remembered.

> Stories were legion of how these union leaders tolerated and even promoted drug and alcohol use in the Fremont factory before it was NUMMI, and how workers sabotaged cars to guarantee overtime pay fixing them in the factory yards. Absenteeism was rampant because workers hated their jobs and management.‡

Both joint venture partners had to be fully aware of the workforce challenges. Despite these challenges, the joint venture parties agreed to rehire the old GM workforce. When the joint venture began, management and the union sent 5,000 job applications to former employees. They received 3,000 replies and re-hired 85%. "Of those who applied, many voluntarily dropped out during the screening process when they learned what would be expected of them in the new operation. Only 300 applicants were rejected, mostly for poor work histories or drug or alcohol problems."§

Why Did Toyota and GM Agree to This Joint Venture?

Despite all the challenges, or perhaps because of them, GM and Toyota reopened the plant as a joint venture in 1984 to manufacture vehicles to be sold under both brands. GM saw the joint venture as an opportunity to learn about TPS, while Toyota wanted to see if TPS could be implemented successfully outside of Japan.

---

* Langfitt.
† Langfitt.
‡ Kiley, D. Goodbye, NUMMI: How a plant changed the culture of car-making. April 2, 2010. Available at www.popularmechanics.com/cars/a5514/4350856/.
§ O'Reilly and Pfeffer, pp. 184–185.

GM "had two major objectives in entering the joint venture: 'to gain first-hand experience with the extremely efficient and cost-effective TPS' and to obtain high-quality automobiles for its Chevrolet division (Community Relations Department, 1990). GM hoped that it could apply what it learned at NUMMI in its other plants, and thus gain great benefits company-wide."*

THE PRINCIPLES IN ACTION

While it is impossible to go back in time and visit NUMMI, there are many reports that describe the systems, tools, behaviors, and culture. These reports allow the reader to use this case study as a way to prepare to "go and observe" the principles of enterprise excellence in the *Shingo Model* in action.

Cultural Enablers: Respect Every Individual

Respect must become something that is deeply felt for and by every person in an organization. Respect for every individual naturally includes respect for customers, suppliers, the community, and society in general. Individuals are energized when this type of respect is demonstrated. Most associates will say that to be respected is the most important thing they want from their employment. When people feel respected, they give far more than their hands—they give their minds and hearts as well.

To better understand the principle of respect for every individual simply ask the question "why"? The answer is because everyone is a human being with worth and potential. Because this is true, every individual deserves respect.

At NUMMI, respect started during the hiring process. GM Manager Jamie Hresko, who went through the hiring process as a new recruit to see how it worked, described it as follows:

> I had the opportunity to go through the hiring process, including written tests, interviews with other hourly people, and the training program. I was amazed at all the levels of testing and educational training that they provide new recruits. ... The initial training and orientation program takes three months ... The formal training is 30 days and includes aerobics, instruction

---

about the Toyota Production System, how the suggestion system operates,
the importance of standardized work processes, scrolling, welding, the team
process, and lots of discussion of the importance of having the right attitude.
It gave a very realistic picture of what it's like to work on the line. NUMMI
doesn't want people unless they have the right attitude and are capable
of performing the aggressive work requirements. They have to fit with the
company.[*]

Respect was continually shown, especially to line operators, through
cross-training, consideration for ergonomic impact on operators, and
safety. Those in a team rotated jobs every couple of hours to share the bur-
den and provide ergonomic relief. Team members understood that if the
job was too difficult or unsafe they would redesign the work.[†]

As Hresko described it:

Everyone seemed to agree that all team members must be on board if you're
going to survive and compete in today's competitive market. There was a
sense that the system wasn't designed to squeeze people or destroy them but
to help them be competitive. The peer pressure is intense. Most people were
willing to stay after their shift to finish their job if necessary. I wasn't accus-
tomed to this type of attitude. People at NUMMI just go do it. They believe
that their job is to protect the customer by never shipping a bad product. But
this isn't because of a fear of being fired. It's actually appears to be harder to
fire people at NUMMI.[‡]

Training did not stop once a new employee was onboarded, but continued throughout the time they were at NUMMI. When team members
were hired, they attended a four-day orientation led by team members and
managers. "There were classes on the team concept, the Toyota Production
System, quality principles, attendance requirements, safety policies, labor
management philosophies, cultural diversity, and the competitive situation in the automobile industry."[§]

With the Toyota system, there was continual training. One plant manager said that people were given the opportunity to develop their skills,
and they received adequate training for advancement and promotions.
"The approach is characterized as a 'lifetime training system'. For example,

---

[*] O'Reilly and Pfeffer, p. 179.
[†] O'Reilly and Pfeffer, p. 180.
[‡] O'Reilly and Pfeffer, p. 180.
[§] O'Reilly and Pfeffer, p. 189.

a manager noted, 'group leaders should be a kind of an instructor—not a commander. You can't do this without training.'"*

Newly promoted supervisors received 13 weeks of training in how to treat people, and periodic special training sessions were designed and offered to support annual goals, in addition to the ongoing training programs in problem-solving, creativity, quality improvement, industrial engineering, and leadership. A program to help employees earn their high school diplomas was also created. Martha Quesada, a team member, reflected on the dominant attitude that "the company feels that the more employees know, the more well-rounded they are, ... the more valuable they are to the company."†

Respect for people extended well beyond training, and into all aspects of the NUMMI experience. People were clearly the foundation for the NUMMI production system. It was felt that the company was only as good as its people, individually and collectively; therefore, the primary responsibility of management and staff was to support the production people.

The *NUMMI Team Handbook* stated:

> *Our HR philosophy guides us in the development of our full human potential to enable us to build the highest quality automobiles at the lowest possible cost by:*
>
> * *Recognizing our worth and dignity*
> * *Developing our individual performance*
> * *Developing our team performance*
> * *Improving our work environment*‡

To foster mutual trust and respect, equity, involvement, and teamwork, NUMMI instituted and practiced the following: "Job security (a no-layoff policy), concern for safety in the plant, individual responsibility for quality, active involvement in the decision-making process, no time clocks, common eating and parking areas, and no distinctions in dress."§ Several core values were continually strengthened through "customer satisfaction (quality and cost), dignity, trust, continual improvement, simplicity, and harmony."¶

---

* O'Reilly and Pfeffer, p. 193.
† O'Reilly and Pfeffer, pp. 193–194.
‡ O'Reilly and Pfeffer, p. 188.
§ O'Reilly and Pfeffer, pp. 188–189.
¶ O'Reilly and Pfeffer, pp. 188–189.

Respect for employees was also manifest in how the operators were trusted and empowered. Under the NUMMI system, all workers were responsible for "quality and safety and provide[d] a method (the *andon* cord) for any person to stop the line to get help with a quality or safety problem—even though the estimated cost of line downtime [was] $15,000 per minute. The cord [was] routinely pulled over 100 times per day.* One person said:

> *Any worker on the line can stop production if they see a problem.... They're actually required to do so. Why? Well, because quality depends on it, and the survival of our company depends on our quality. Only because of this level of involvement on the part of every employee were we able to take #1 on the J.D. Power survey.**

Authors O'Reilly and Pfeffer said, "This approach is very different from that used in the typical U.S. manufacturing operation and requires different management skills. One NUMMI manager noted that for the line to work efficiently, managers must respect their employees:

> *One of the key concepts is respect for the worker, the team member. The Japanese know that to make things more waste-free and streamlined, they have to work with the people on the line. They have to work with their people, to listen to them for ideas, and to work with them to support theirs ... They trust their team members are doing their best."†*

Cultural Enablers: Lead with Humility

One common trait among leading practitioners of enterprise excellence is a sense of humility. Humility is an enabling principle that precedes learning and improvement. A leader's willingness to seek input, listen carefully, and continuously learn creates an environment where associates feel respected and energized and give freely of their creative abilities. Improvement is only possible when people are willing to acknowledge their vulnerability and abandon bias and prejudice in their pursuit of a better way.

The right kind of culture cannot be developed without proper leadership. From the beginning, management at NUMMI led with the intent to create a culture of serving the needs of operators.

---

* O'Reilly and Pfeffer, pp. 191–192.
† O'Reilly and Pfeffer, pp. 190–191.

Gary Convis, a former executive vice president at NUMMI, was cautioned by Kan Higashi (then NUMMI's president), when first promoted as the vice president of manufacturing, to manage NUMMI's manufacturing operations as if he had "no power." Higashi said, "Everyone knows you are the vice president; however, in your day-to-day job, your listening, coaching, mentoring, and gaining consensus around key initiatives will be most appreciated and effective."*

Higashi stressed:

> *Toyota understood the importance of using all means available to establish a climate of fairness—including fewer levels of management, no executive perks, and a blurring of distinctions between managers and team members. ... Higashi noted that this approach was designed at NUMMI to signal their vision and intent. Kosuke Ikebuchi, NUMMI's first head of manufacturing, used to emphasize to his managers, "Never forget, management is the beneficiary of our team members' hard work. Our job is to support their efforts. This way the company will be successful."*†

Jamie Hresko, the GM manager that went through the hiring process as a new recruit, described this type of leadership:

> *I have always believed that supporting the hourly technicians is the most important factor in winning in the automobile industry. I was overwhelmed by how much the NUMMI process is geared toward helping line workers. Here, production really is king. As a team member, you can always get engineering help. The goal is to support the operator. I feel that establishing processes and systems to engage hourly technicians and team support is the key to success.*‡

Leading with humility was a key in building the right kind of culture at NUMMI.

Continuous Improvement: Seek Perfection

Perfection is an aspiration not likely to be achieved but the pursuit of which creates a mindset and culture of continuous improvement. The realization of what is possible is only limited by the paradigms through

---

* O'Reilly and Pfeffer, p. 191.
† O'Reilly and Pfeffer, p. 191.
‡ O'Reilly and Pfeffer, pp. 181–182.

which one sees and understands the world. The concept of "continuous improvement" stems from the Japanese word *kaizen*. The effort to improve should involve everyone in an organization, encompassing all functions in the organization, constantly seeking ways to improve.

Masaaki Imai, the kaizen expert, defines kaizen in the following way:

> *Kaizen means ... "improvement." The word implies improvement that involves everyone—both manager and workers—and entails relatively little expense. The kaizen philosophy assumes that our way of life ... should focus on constant improvement efforts. ... The kaizen process brings about dramatic results over time...*
>
> *Most "uniquely Japanese" management practices, such as total quality control (TQC), or company-wide quality control and quality circles, and our style of labor relations, can be reduced to one word: kaizen. Using the kaizen in place of such buzzwords as productivity, total quality control (TQC), zero defects (ZDs), just-in-time (JIT), and the suggestion system paints a clearer picture of what has been going on in Japanese industry. Kaizen is an umbrella concept for all these practices.**

An important part of the kaizen philosophy is the systematic elimination of waste. TPS commonly refers to seven wastes:

- Overproduction
- Inventory
- Defects
- Motion
- Processing
- Waiting
- Transport†

NUMMI adopted kaizen in the way Imai described. It was the overarching philosophy behind the design of the company's management systems. Authors O'Reilly and Pfeffer summarize how the NUMMI production system drove kaizen throughout the production process.

> *Obviously, the NUMMI production system is characterized by a constant tension and quest for improvement. There is always a danger, on one side,*

---

* Imai, M. *Gemba Kaizen: A Commonsense Approach to a Continuous Improvement Strategy*, 2nd edition. New York, NY: McGraw-Hill, 2012, pp. 1–2.

† Imai, pp. 80–87. This is an excellent summary of the seven wastes.

of becoming complacent or, on the other side, of reverting to old top-down methods of driving production. Balancing this tension requires consistency in adhering to the principles of the system and requires maintaining a level of cooperation and trust between management and employees. Trust of this sort is indispensable in the NUMMI system. And building this trust requires a genuinely open, data-driven, decision-making management style, but not necessarily one that is democratic or permissive. It was one of disciplined analysis and constant questioning and listening, not one of speed and individualism.

Managing this process also requires continual change. To do this effectively over long periods of time requires that managers continually "renew the spirit," as one manager put it—that is, find new ways to reenergize employees to push for improvement and to avoid the complacency that success brings. These efforts involve the usual buttons and banners but also involve defining new challenges with which to engage the employees' interests and energies.

The results aren't perfect. Employees always acknowledge that it is an ongoing effort. Says team member Martha Quesada: "It's not always a honeymoon. We still have a lot of problems and there are still some conflicts, but we're working constantly to keep improving. We have a foundation for communication and teamwork, and that's what's important."

*NUMMI managers and employees recognize that the quest for continuous improvement is never over. The answer isn't in some high-tech solution but in the people. Bill Childs, vice president of human resources, believed that "It all centers around the treatment of people and the dignity you give the hourly person on the line."**

People are attracted to excellence. They want to be part of something that is truly best-in-class. The "quest" for perfection inspires people to keep getting better and better, and to make their workplaces better and better.

Continuous Improvement: Embrace Scientific Thinking

Innovation and improvement are the consequence of repeated cycles of experimentation, direct observation, and learning. A relentless and systematic exploration of new ideas, including failures, enables one to constantly refine their understanding of reality. The scientific method demands that any improvement must be compared to a standard. Otherwise, it is not clear if an improvement is really an improvement. This is often referred to as "standard work."

---

* O'Reilly and Pfeffer, pp. 195–196.

An example of how the scientific method and standard work drove improvements at NUMMI includes team members being responsible to design and improve their jobs, including the industrial engineering, generating detailed definitions and sequencing of jobs, completing standardized work sheets, and adhering to these instructions. One NUMMI employee said, "We're responsible for timing our own jobs. We're always involved in making changes—not to add tasks but to improve safety and efficiency."*

A NUMMI manager described the impact of standard work in this way:

> *We get improved quality because workers identify the most effective procedure for the job. When you have a good procedure, any problems with equipment very quickly come to the surface. And since every worker becomes a real expert, that means that each workstation becomes an inspection station. ...*
>
> *Standardized work also means that each worker in the team can refer to a good procedure for doing the job, so that even if ... one team member is absent, quality doesn't go through the floor. In the old days, absences killed quality because the replacement not only didn't know the job but didn't even have a procedure to refer to for doing it right. When you've got standardized work, you've got a clear base on which you can build to make continual improvements—you can't improve a process you don't understand. And standardized work has the major benefit of giving control of the job to the person who knows it best—it empowers our workers.*
>
> *So you see that standardized work ties together a lot of different elements of the NUMMI production system. It's kind of the foundation of the whole thing.†*

Following standard work and the scientific method was not just practiced on the individual level, or the team level, but extended throughout the plant. NUMMI made performance highly visible in order to increase the employees' sense of interdependence and teamwork. Standardized work and kaizen charts were hung in public team areas, and attendance boards with individual ratings and defect records were displayed prominently throughout the plant. It was the team's responsibility to maintain these records. Daily meetings were held with 40–50 assistant

---

* O'Reilly and Pfeffer, p. 192.

† Adler, P. S. The learning bureaucracy: New United Motors Manufacturing, Inc. In B. M. Staw and L. L. Cummings (eds.), *Research in Organizational Behavior*, Greenwich, CT: JAI Press, April 1992, p. 38. Available at www.bcf.usc.edu/~padler/research/NUMMI(ROB)-1.pdf.

managers and team and group leaders to discuss defects found in randomly selected cars. Managers were asked to explain the defects and describe the corrective measures they would take, all in a blame-free spirit with a focus on solving the problem instead of holding an individual responsible. Team leaders would then pass this information on to team members so all employees understood how the plant was performing on a daily basis.*

Constantly generating ideas and testing them against the standard work allows individuals, teams, and the entire organization to see if the ideas generated improve quality, safety, and productivity.

Continuous Improvement: Focus on Process

All outcomes are the consequence of a process. It is nearly impossible for even good people to consistently produce ideal results with a poor process both inside and outside the organization. There is a natural tendency to blame the people involved when something goes wrong or is less than ideal, when in reality the vast majority of the time the issue is rooted in an imperfect process, not the people.

This is a principle that is often better understood in the context of how it interrelates with other principles. This is demonstrated by reports from NUMMI, such as the "blame-free spirit" mentioned above. The continuation of a quote cited earlier implies this focus on process combined with respect for every individual.

> *They trust their team members are doing their best. When something breaks down, managers feel it's their responsibility and they're apologetic out of respect for their team members.†*

When the process results in poor quality, or a safety hazard to an operator, it is usually due to a faulty process. Managers feel "responsible" because they are responsible for designing the process that has failed. The focus on process principle brings attention to improving the process, so the problem can be solved in the present and prevented in the future. It usually requires involvement of the operator or associate because they tend to know the process better than anyone else. And it

* O'Reilly and Pfeffer, p. 194.
† O'Reilly and Pfeffer, p. 191.

often involves going to the actual workplace and observing the actual work being done.

Expectations of group leaders and supervisors at NUMMI also demonstrate this focus on process. Under supervision of the team leader, team members were even responsible for defining standard work and for measuring and setting their own time standards.

> *In Japan, even though everyone is trained and understands standardized work, it's usually the Team Leader that performs the standardized work analysis. And that seems to work out OK because of the level of trust they have with their team members. Here at NUMMI, because of the history of more conflictual relations with supervisors and industrial engineers, it's really important that workers perform the analysis themselves.**

This self-reliance on the team at NUMMI extended much further into assisting in the resolution of engineering problems and breakdowns, implementing the suggestion system, solving problems, and supervising any corrective discipline or counseling issues. Team leaders also helped out on the line as needed.† Second-level managers (supervisors) ... "are expected to be more process oriented than results oriented. In addition to responsibility for budgeting, planning, and training, managers are expected to encourage openness concerning problems, and to see that problems are resolved at the lowest possible level."‡

Continuous Improvement: Assure Quality at the Source

Perfect quality can be achieved only when every element of work is done right the first time. If an error should occur, it must be detected and corrected at the point and time of its creation. The NUMMI story is filled with this principle in action.

> *The responsibility for quality does not reside in management supervision and inspections but is pushed down to the worker under the principle of ... identifying problems when they occur, and responding to them immediately. Problems are resolved at the lowest level possible. The assembly line is kept constantly alert, with the emphasis being on doing the operation*

---

* Adler, p. 36.
† See, for example, O'Reilly and Pfeffer, p. 190.
‡ O'Reilly and Pfeffer, p. 190.

correctly every time. Responsibility lies with the individual to call atten-
tion to a problem whenever a defect is observed. This places a premium on
people being able to identify problems and to quickly adjust and to correct
errors.*

Workers try to correct the problem on the line. If it takes too long to fix, the
line stops. The andon cord also plays a surprisingly cheerful little song that
workers can choose. For longtime GM workers who switched to the NUMMI
system, all this was a revelation. When Rick Madrid trained in Japan, he saw
workers stop the line to fix a bolt. "That impressed me," he said. "I said, 'Gee
that makes sense.' Fix it now so you don't have to go through all this stuff.
That's when it dawned on me. We can do it. One bolt. One bolt changed my
attitude."†

As is made clear, the andon cord is not only a way of empowering and
trusting people, but it is also a valuable quality control tool. This culture
permeated the operators: "They believe that their job is to protect the customer by never shipping a bad product."‡

Jamie Hresko tested to see what would really happen if he missed some
quality checks. He reports: "When I missed a couple of quality checks an
operator down the line picked them up and stopped by to make sure I
didn't do it again."§

In an article celebrating NUMMI's 20-year anniversary, Edwin Duerr
and Mitsuko Duerr of San Francisco State University said:

The company's commitment to quality was clearly illustrated in August 1990.
It was discovered that parts that had been received from a new supplier
were defective. Rather than continue production with parts that might later
require replacement, the plant was shut down for three days until new parts
could be obtained. Cars that had already been produced were not shipped
to dealers but were held for part replacements. Since it was not the workers
fault that the parts were defective, and NUMMI wanted to encourage them
to report defects, the company offered the workers full pay for the period the
plant was shut down.¶

As quoted above, a worker said: "Any worker on the line can stop production if they see a problem…. They're actually required to do so. Why?

---

* O'Reilly and Pfeffer, p. 192.
† Langfitt.
‡ O'Reilly and Pfeffer, p. 180.
§ O'Reilly and Pfeffer, p. 180.
¶ Duerr and Duerr, pp. 8–9.

Well, because quality depends on it, and the survival of our company depends on our quality."*

Assuring quality at the source is the only way to see quality problems where they occur, and to see them reliably. Inspectors viewing problems later in the process do not know the causes of problems, and are not engaged in preventing those problems from recurring.

Continuous Improvement: Flow & Pull Value

Value for customers is maximized when it is created in response to real demand and a continuous and uninterrupted flow. Although one-piece flow is the ideal, often demand is distorted between and within organizations. Waste is anything that disrupts the continuous flow of value.

> *In pull production, all processes should be arranged so that the work-piece flows through the workstations in the order in which the processes take place. … Once the line is formed, the next step is to start a one-piece flow, allowing only one piece at a time to flow from process to process. This shortens lead time and makes it difficult for the line to build up inventory between processes.†*

A pull production system requires that all problems—quality, machine breakdowns, etc.—be addressed before the process can be re-started. As such, a line with good flow quickly forces the identification and resolution of problems.

NUMMI adopted the pull production systems in place in other Toyota plants. These production systems are often referred to as JIT and *kanban*.

> *The just-in-time (JIT) inventory system is designed to produce only what is being ordered or sold rather than to produce for inventory that will be used to absorb ups and downs in demand. Lowered (or eliminated) inventories of incoming, in-process, and finished goods saves space and costs of money tied up. It also:*
>
> - *Results in quicker identification of problems arising due to defective inputs or processing problems;*
> - *Results in increased emphasis on avoiding breakdowns (and thus on preventive maintenance); and*

---

* O'Reilly and Pfeffer, pp. 191–192.
† Imai, pp. 158–159.

- *Provides additional pressure to make production processes more flexible, such as being able to produce more types and styles of vehicles on one assembly line (as is done at NUMMI).**

Many of the other principles already discussed, such as seek perfection (especially through the elimination of waste), focus on process, and assure quality at the source, all result in a smoother flow of product, thereby providing more value for the cost.

Enterprise Alignment: Think Systemically

Through understanding the relationships and interconnectedness within a system, the reader is able to make better decisions and improvements.

The NUMMI Production System (NPS) can only be understood by looking at how the components are integrated together as essential pieces of a whole system. As reported:

> *The most frequent explanation offered by cynics who had never worked at the plant for this turnaround was that the laid-off workers would do almost anything to get their jobs back. But knowledgeable insiders discounted this explanation. After all, the threat of plant closings has done little to enhance performance at other GM plants. ... The success at NUMMI comes from an integrated set of HR and manufacturing processes that align the interests of employees, managers, and the company and involve the workforce in a way that simultaneously empowers them while managing the interdependence inherent in a complex manufacturing process.†*

The company reinforced this integrated approach through their reward system. The NUMMI system relied on a flat wage structure to balance rewards and nurture fairness. Incremental rewards were not given to learn a new skill or task, which helped support the belief that the company's success depended on everyone's efforts. Some U.S. managers have a hard time grasping this concept, because they believe that employees will only work hard if they are given incremental monetary rewards. One NUMMI senior manager said:

> *Our team members are ready and willing to change as long as they feel they are being treated fairly and equitably. We've tried to avoid favoritism and to level out the harder jobs. A single-pay level is fundamental to the success of*

---

* Duerr and Duerr, p. 10.
† O'Reilly and Pfeffer, pp. 186–187.

this company as is security of employment. We have learned … the impor-
tance of tying the company's success, and the success of the individual, to
*things they can control.**

NUMMI helped enhance the ability of their employees to think about
the entire system through cross-training and job rotations, and through
the plant-wide, visible displays of plant performance. Rather than call the
system TPS, it is impressive that NUMMI called their system the NUMMI
Production System (NPS). This gave employees a sense that it belonged to
them and that they could change and improve it as it fit their needs.

Enterprise Alignment: Create Constancy of Purpose

An unwavering clarity of why the organization exists, where it is going,
and how it will get there enables people to align their actions, as well as to
innovate, adapt, and take risks with greater confidence.

From the beginning of NUMMI, management created a clear sense of
purpose. This is described in the following:

> *The primary goal of the Toyota manufacturing system is to reduce costs and*
> *maximize profits through the systematic identification of waste. At NUMMI,*
> *this goal was broadened from reducing the cost per vehicle to include continu-*
> *ally improving quality and securing safety.* The vision for NUMMI was to
> produce the highest quality, lowest cost vehicles in the world.†

As seen from the numerous examples provided, employees at all levels of
the organization rallied around this vision and worked hard to achieve it.

Results: Create Value for the Customer

Ultimately, value must be defined through the lens of what a customer
wants and is willing to pay. Organizations that fail to deliver both effectively and efficiently on this most fundamental outcome cannot be sustained over the long term.

> *When GM closed the plant, the Fremont plant was the worst plant in the*
> *GM system. After a few years of operation, NUMMI was the best plant in*

---

* O'Reilly and Pfeffer, pp. 192–193.
† O'Reilly and Pfeffer, p. 185; emphasis added.

the GM system. NUMMI vehicles have won a number of J.D. Power and Associates Initial Quality awards, Top Car under $15,000 ratings from the American Automobile Association, Consumer Digest's "Best Buy" rankings, and other recognition. The company has received a number of J.D. Power and Associates Plant Awards for its factory, and it has received DNV Certification Inc. Environment Management Certification (ISO 14001).*

NUMMI was consistently ranked as one of the efficient producers of automobiles. "The company employed 2.62 workers per vehicle produced in 1996 compared with GM's 3.62—a 20% productivity advantage ... and produced automobiles of the highest quality—approximately 80 defects per 100 cars in 1996 compared with around 110 for GM."[†]

LESSONS LEARNED

Lesson #1: TPS, or Lean Manufacturing, Works Anywhere in the World

Kan Higashi, the second president of NUMMI, recalled that at first Toyota was concerned that American workers and the UAW would not understand the Toyota production concepts. But, he said, "We found people here to be capable and flexible" and he didn't see much difference between American and Japanese employees. He noted that management treats people not as part of a machine but as human beings deserving trust and respect. The result? "Basically the NUMMI plant is the same as the plant in Japan—only smaller."[‡]

When NUMMI started showing promising success, Toyota invested in its Georgetown, Kentucky plant.

The diversity of the workforce, the various problems within the workforce prior to NUMMI, and the union, represented a worst-case situation. As far as can be determined, it seems Toyota decided to use NUMMI as a test, because NUMMI had more problems than Toyota could experience anywhere in the world. In other words, if NUMMI could work, then Toyota could safely expand to anywhere in the world and make TPS work there too.

---

* Duerr and Duerr, p. 2.
[†] O'Reilly and Pfeffer, p. 185.
[‡] O'Reilly and Pfeffer, p. 184.

Lesson #2: Lean Tools and Systems Do Not Thrive without the Right Culture

David Kiley, author of the article "Goodbye, NUMMI: How a Plant Changed the Culture of Car-Making," said:

Some have suggested that because GM still lags Toyota on quality in North America, the NUMMI experiment failed. Author Jeffrey Liker (The Toyota Way, McGraw-Hill) says that what makes the deal look bad in retrospect was the poor way in which GM adopted what its managers learned from the Toyota Production System.

While today's "Global Manufacturing System" manual at GM is a direct copy of the Toyota Production System, Liker says, "It took fifteen years for GM to take the lessons learned at NUMMI seriously." And after it began the process, it took another five years before GM really started to see substantial impact on its overall system in things such as higher quality scores and productivity gains.

Cultural arrogance and lack of focus were at the heart of why it took so long. GM's Roger Smith launched the Saturn division in large part to provide GM with a blank canvas on which it could start a new business model within the company that would hopefully serve as an internal university. The work rules and union agreement at Saturn's Spring Hill, Tenn., plant were modeled on the NUMMI system. GM had seen how some of the worst rated workers in the country, from the previous operation in Fremont, could be transformed under the right system—namely, Toyota's. Flexible work rules were adopted. "Team" was at the heart of the Saturn mantra: empowering workers to stop the production line if a problem needed to be fixed; rewarding workers for solving problems in production and adding to quality. These were all tenets that Toyota brought to NUMMI and that GM NUMMI managers took to Spring Hill—and that managers and workers took from Spring Hill to other GM plants. But it was a slow process.

Would the work-rules changes and quality improvements now embedded in GM have been likely if not for NUMMI? They might have taken hold under the sheer weight of pressure from Toyota and Honda and the growing influence of public quality ratings from Consumer Reports and J.D. Power and Associates. But we'll never know for sure.*

In his book, *The Toyota Way*, Professor Liker was even more specific. After describing GM's implementation of the work group structure and some of the TPS tools, he writes:

---

* Kiley.

What GM was lacking was obvious: it did not have the Toyota Production System or the supporting culture. It merely copied and appended the work group structure onto traditional mass-production plants. The lesson was clear: don't implement work teams before you do the hard work of implementing the system and culture *to support them.**

A more detailed explanation of the struggles GM faced in their attempts to adopt the learnings from NUMMI are explained by Duerr and Duerr:

General Motors' two objectives were to gain experience with the Toyota Production System and to obtain high quality automobiles for its Chevrolet Division. General Motors did gain valuable experience with the New United Motor Manufacturing, Inc. joint venture, but found it difficult to apply what it had learned to other GM plants. They also obtained high-quality small cars but, because of the image and marketing problems discussed above, were not able to sell as many of them as they had expected.

General Motors provided a number of managers with experience in working at NUMMI, and thousands of workers and managers with visits to the Fremont facility (Moore, 1988, 35). But this experience did not prove as valuable as had been hoped for several reasons. One factor was that in their assignments following work at NUMMI, the managers were not kept together, but rather distributed to various positions around the company. In their post-NUMMI positions they were surrounded by workers and managers whose traditional adversarial relationships were so entrenched that they simply could not be changed by one individual or even a small group of people (Interview, 1990a). It could reasonably be argued that the adversarial relationships were so strong that even as a group they could not have changed the culture of a single existing plant.

Experience with NUMMI was also applied by GM in an innovative small-car project named Saturn, originally conceived by GM CEO Roger Smith in 1983. It was to "make superb little cars to beat the Japanese at their own game" (Taylor, III, 2004, 119). In order to free the project from the bureaucratic constraints of GM, allow the development of cooperative labor management relations, and to provide it with its own identity, it was set up as a separate company. Any GM employees desiring to work for Saturn had to give up their positions and seniority with GM. After seeing the results achieved at NUMMI with just an average level of automation, GM scaled back the level of automation to be used at Saturn.

Saturn was a success in achieving a high level of labor management cooperation, gaining sales to people who previously had purchased Japanese or

---

* Liker, J. K. *The Toyota Way*. New York, NY: McGraw-Hill, 2004, p. 185; emphasis added.

European cars (70% of first time Saturn buyers had previously owned for-eign-nameplate automobiles), and achieving high customer loyalty. However, most Saturn buyers never traded up to other GM cars, and looked to other brands when they wanted bigger or more stylish automobiles (Welch, 2002, 80; Taylor, III, 2004, 125). Saturn developed right-hand drive cars for export to Japan in 1997, but disappointing sales resulted in abandoning the effort in 2000 (Taylor, III, 126). The company has never made money in its history. ...

Though it did learn and apply much from the Toyota Production System, it was unable to replicate the system in any existing plant. Workers and man-agers at existing GM plants have such a long history of confrontational rela-tions, and such a distrust of each other, that a system based on mutual trust and cooperation apparently could not be implemented.*

Lesson #3: It Is Possible to Turn Around Any Workforce by Applying These Principles

Another remarkable aspect of the NUMMI story is that these principles are effective in turning-around poor performing organizations. The GM Fremont plant was the worst in the GM system. Why is it that NUMMI was able to achieve such great results in such a short period of time?

It's not because it had some special workforce. The plant employed ex-GM workers. It's not the technology. From its inception, NUMMI relied on older technology and was not as automated as many competitive plants. It's not that NUMMI is a nonunion workplace, because the company had the same union (and originally even the same union leaders) as at the old GM-Fremont plant. How was NUMMI able to continually produce with such high levels of quality and efficiency when they began by reopening a plant that was prob-ably one of the worst in the GM system at the time it was closed in 1992? ... How can one management achieve such extraordinary results when previous management failed miserably with the same people?[†]

The explanation for the mystery of NUMMI's success rests on something far more subtle—the values of trust, respect, and continuous improvement that characterized relations within the plant, and the consistency with which those were applied in all the operating systems and management practices. This consistency in alignment was manifest in how people were selected, trained, rewarded, and supervised. But this explanation probably seems sim-plistic and unsatisfying. ...

---

* Duerr and Duerr, p. 12.
† O'Reilly and Pfeffer, p. 177; *Note:* Tenses have been adjusted from present tense to past tense.

*This system is so difficult to imitate because of something fundamental and not very amenable to change, namely, the basic assumptions that management begins with in developing relationships with employees. At NUMMI ... the fundamental belief on the part of management was that people are responsible and want to contribute. Unleashing their potential requires that they be treated accordingly.**

CONCLUSION

The purpose of including this case study is to help the reader "observe" ideal behaviors, and how ideal behaviors lead to excellent results. These behaviors were driven by a clear and constant sense of purpose—"to produce the highest quality, lowest cost vehicles in the world." These behaviors were also driven by the overarching system—NPS—as well as many supporting systems, such as training, compensation, recognition, no layoffs, quality control, visual management, etc. And, most importantly, the behaviors are clearly grounded in the principles of operational excellence.

The challenge for the reader is to "go and observe" behaviors in their own organizations to see if those behaviors are guided by a clear sense of purpose. Are systems driving ideal behaviors, or less than ideal behaviors? Are behaviors grounded on the principles of operational excellence? If gaps are observed, then how can those gaps be closed?

In discussing the NUMMI case with several groups around the world, the two most common questions asked are (1) "If NUMMI was so great, then why did it get closed down?" and (2) "That may be great for manufacturing, but can Lean work in nonmanufacturing organizations?"

As for the first question, there were a number of factors that led to the closing. Both parties in the joint venture had accomplished their initial objectives. GM had many years to learn TPS first-hand. Toyota learned that TPS works in a multicultural setting in countries outside of Japan. Neither party redefined their original objectives after these objectives were completed, so the plant was already ripe for closure before the economic downturn in 2008. Owing to a declining market, and the great recession of 2008, GM was forced into bankruptcy. As part of the restructuring plan, GM announced in 2009 that they were pulling out of the joint venture.

---

* O'Reilly and Pfeffer, pp. 196–197; *Note*: Tenses have been adjusted from present tense to past tense.

With half of the production volume siphoned away by GM, and absent a joint venture partner, Toyota closed the plant in 2010, and transferred ownership to Tesla. It is now the "Tesla Factory."

As for the second question, applying the principles to drive operational excellence in industries besides manufacturing organizations is well documented in healthcare, financial institutions, processing organizations, service organizations, and so on. Applying the principles to drive operational excellence in functions besides the manufacturing functions—such as accounting, HR, IT, etc.—is also well documented. Case studies representative of how the principles can be applied in nonmanufacturing organizations and in many different functions are recorded in publications that have received the Shingo Publication Award.*

---

* See http://www.shingo.org/publicationaward.

10

Go and Observe

Constantly think about how you could be doing things better, and keep questioning yourself.[*]

Elon Musk

Co-founder and CEO, Tesla Inc.

THE GEMBA WALK

The reader may wonder why an entire chapter, even though it is a short one, should be dedicated to the concept of the gemba walk. It is because this is a critical process in transforming an organization toward the *Shingo Guiding Principles*. As learned from Shigeo Shingo and his son Ritsuo, without the gemba walk, one wastes a lot of time discussing something they really don't understand.

The Japanese term gemba means, "Go to the actual place where the work is being done. Go to where the problem is." But the application of this tool is much more than just paying the place a visit. "Go and observe" is a more accurate way of explaining what it is that needs to be done. One goes to the place where the work is being done and one quietly sits back and observes. One is not there to interfere with the process, or to make anyone nervous. The job is to watch and learn.

In the words of the master, here is what Ritsuo Shingo has to say about the importance of the gemba:

> *My name is Ritsuo Shingo and I am from Toda city, Japan. I used to work for Toyota motors more than 40 years. My definition of gemba is the place*

---

[*] Ulanoff, L. Elon Musk: Secrets of a Highly Effective Entrepreneur. April 13, 2012. Available at http://mashable.com/2012/04/13/elon-musk-secrets-of-effectiveness/#AMxPBOefMaq5.

where the actual thing is happening. That is gemba. Not necessarily the plant.

Gemba is very important. Almost everything is there. Program, improvement model, many things are there. Without knowing gemba, just being in the office, how can you find the problems or improvement area without knowing gemba? No. Even if you are in the meeting room and calling many people, without watching and seeing gemba in most cases, in my opinion—it doesn't work.

You should go to gemba, where the actual thing is happening. It's the most effective way to grasp the fact, find the problem, and find the improvement area.

Even at Toyota, the manager should know what the gemba is. Knowing is one thing, but actually implementing, doing is another. Even at Toyota, we are losing the good attitude. So even the basic concept of gemba, he doesn't understand. He should have a very big eye—you should watch carefully, and big ear—listen very carefully, but not a big mouth. So, the leader should be a very good listener. If you talk too much to the people, they just listen and they just try to do what you say. No, the opinion from the workers—the gemba people—is the best way to know the fact.

Far too often the translation and interpretation of a gemba walk involves a parade where a group of individuals march down to the place of the failure and ask a lot of questions, which are fielded by the supervisor who often wasn't even involved in the failure. That's not a gemba walk. That's not how a "go and observe" exercise should be executed. So it's time to walk through the "go and observe" process to see how an effective "walk" is executed. But rather than using the "go and observe" to solve a problem, use it to study one of the *Shingo Guiding Principles* and the behaviors associated with that principle. Here is a step-by-step process to consider:

Step 1. Decide what to focus on: Borrowing from the exercise in Chapter 8, look at the charts that were created to demonstrate that "Principles Inform Ideal Behavior." Early in that chapter, the reader built a table to rank their behavior performance against each of the principles. Now the reader can combine the learning from each of these charts. From the performance chart, identify the principles where one feels performance is the worst. Select one for further study and analysis through the "go and observe" process. Then, from the "Principles Inform Ideal Behavior" chart, look at the behaviors that one would expect to observe from this failing principle. For example, use respect every individual as the principle where the enterprise is the weakest (see Figure 10.1).

Principle:	Respect Every Individual
Leaders:	Leaders always acknowledge and recognize the specific behaviors they see that are close to ideal.
Managers:	Every manager makes certain that large group meetings begin with a safety briefing.
Associates:	Associates demonstrate an eagerness to learn new skills, take initiative, and share their learning and success with others.

FIGURE 10.1
Principles Inform Ideal Behavior Example.

Step 2. Start to fill in the sheet shown in Figure 10.2: Across the top of the sheet, identify the area for observation (the physical location in the facility), the focus of the gemba effort (which principle or behavior to work on), and the round (the first gemba walk, vs. the second or the third, etc.). In the upper left-hand block, list the ideal behaviors to study. For this example, look at the behaviors identified in Figure 10.3. The middle left-hand box is used to list the systems the reader thinks will be involved in creating the behaviors one will both observe and hope to find. Across the bottom of the sheet, list a set of questions to answer and identify, with the hope that the answers to these questions will provide understanding of this principle or these behaviors one is looking for. Figure 10.4 shows graphically how the questions should be constructed, focusing first on Tool-Focused questions, which are the easiest for associates to respond to, then moving upward searching for the systems responsible for the behaviors observed using System-Focused questions, and then connecting this to principles by identifying Principle-Focused questions.

It is important to note at this point that when the reader goes to the gemba they will find that all three of these boxes filled out in Figure 10.2 are wrong. They will find that the behaviors they hoped to observe aren't the actual behaviors, and that the systems generating those behaviors are not what was expected. And last of all, the reader will discover that the questions asked were the wrong questions. And

Assessment Notes

SYSTEMS

AREA:	FOCUS:	ROUND:

IDEAL BEHAVIOR

1 _____
2 _____
3 _____
4 _____
5 _____
6 _____
7 _____
8 _____
9 _____
10 _____

KEY QUESTIONS   * Create a series of tool, system and principle questions to ask.

1 _____
2 _____
3 _____
4 _____
5 _____
6 _____
7 _____
8 _____
9 _____
10 _____

FIGURE 10.2
Assessment Notes.

that's why one gemba is never sufficient to get a clear understanding of the enterprise's behaviors.

Step 3. Engage a "go and observe:" Go out and make the observations using the sheet in Figure 10.2. During the "go and observe," the reader will fill out and change everything on the sheet. Learn as much as one can about the set of behaviors studied. In the gemba, don't go as a group and simply stand there and listen to the supervisor. Split up and talk to the actual people doing the work. Stand and

Principle:	Respect Every Individual
Leaders:	Leaders always acknowledge the specific behaviors they see that are close to ideal
Managers:	Every manager makes certain that large group meetings begin with a safety briefing.
Associates:	Associates demonstrate an eagerness to learn new skills, take initiative, and share their learning and success with others

Systems:	Recognition Safety Communication
Tools:	Recognition events and appreciation cards Proper safety equipment Newsletter

FIGURE 10.3
Systems Drive Behavior Example #1.

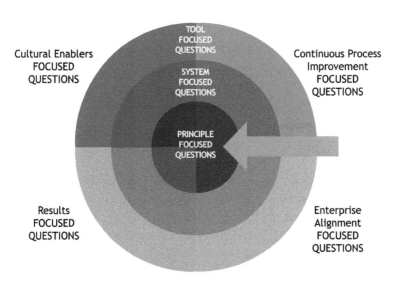

FIGURE 10.4
What Questions to Ask?

observe what's really going on. Seek answers to the questions listed on the sheet. The purpose of questions in the gemba is to help one identify evidence of the actual behaviors of leaders, managers, and associates. If the gemba is less than one hour, then the reader really didn't learn much. A good gemba can run several hours, and it is not unusual to spend half a day simply studying and observing one area.

Step 4. Repeat: Go back as a team and reconvene. Share what was learned. Find out what questions haven't been answered and what new questions should be asked. Tie the behaviors to specific systems. Fill out a new sheet with new questions. Go out and repeat the gemba.

Step 5. Identify the gaps: After one has executed enough gembas to where they feel they have a clear understanding of what the current state is, then it is a good time to reflect on the ideal state and have a conversation around how to get from the current state to the ideal state. What are the gaps? What systems need to be changed? Develop a plan for the execution of those changes. Gain approval to make the changes, and press forward implementing them. Then observe the results. Did the reader get the desired shift in behavior?

Step 6. Learn and try again: Rarely does one make a successful hit the first time through. Working toward the ideal is a step-wise process. It's a trial-and-error process. There is no manual that gives the correct answer to every problem. Try, learn, and try again, each time hopefully moving a little closer to the ideal.

CASE EXAMPLE AND EXERCISE

At this point, it would be valuable to put pen to paper. Photocopy Figure 10.2 and blow it up to an 8.5 × 11″ sheet of paper. Select one or two principles as a focus area. Next, go through an exercise using the case study at the end of this chapter (Vale Clydach Refinery Case Study, a 2014 Shingo Silver Medallion Recipient—see full list of recipients in the Appendix) and look for behaviors and systems that connect with the focused principle. Follow this procedure:

1. Read through the case and identify relevant behaviors and systems.

2. Work through Figure 10.2 sheet(s) to identify ideal systems and ideal behaviors, and think of questions to ask Clydach Refinery employees, just as if the reader was going on the gemba ("go and observe").

3. Go through the case again and see what additional insight on behaviors and systems were gained during the second reading.

4. Consider the questions at the end of the case and put some thought into potential answers to these questions.

5. Think about how this exercise can be applied to one's organization and work out an approach to execute a "go and observe" with a focus on identifying cultural gaps with the opportunity for improvement.

SOME ADDITIONAL "GO AND OBSERVE" TIPS

The reader has already received a long list of suggestions on how to make the "go and observe" experience successful, but here are a few more. When one goes out on the gemba,

- Go out as individuals, not as a team—split up the team
- Get to the associates—don't get pulled away by supervisors and managers
- Start a conversation that connects with the associate
  - "Hi, I'm Bob," or "Hi I'm Susan, do you mind if we ask you a few questions about the work you do?"
  - "What's your title or role?"
  - "So what do you do here?" (break the ice)
- Slowly work into the questions that the reader is there to ask
  - "What do you do when...?"
  - "What happens next?"
  - "Can you tell me more about...?"
  - "How did that make you feel?"

When taking notes, make sure the notes indicate what was observed. But don't stand there and write in front of the interviewee. That makes them feel disconnected. It is important to maintain a social connection.

Make sure to take notes about:

- What work systems were observed?
- What evidence was seen?
- What level was the behavior associated with these systems?
- What were some key strengths?
- What opportunities exist that might allow the enterprise to move to the next level in improving this set of behaviors?

SUMMARY

Figure 10.5 offers a second example that can be used for a "go and observe" exercise. However, offering all the principles and potential behaviors in one figure for the reader would triple the size of the book. Additionally, even if the reader had that complete figure, it wouldn't fit the characteristics of their organization. There is no textbook answer to the behaviors needed to adjust within an enterprise. The reader needs to work those through for themselves. The best the Shingo Institute can do is to wish the reader good luck and offer support when needed.

Principle:	Quality at the Source
Leaders:	Reward identification of defects.
Managers:	Track failures and identify root-cause solutions.
Associates:	Accurately differentiate between good and bad parts.

Systems:	Recognition Problem-solving Visual management
Tools:	Recognition for errors caught Visual verification of part quality

FIGURE 10.5

Systems Drive Behavior Example #2.

We can only learn from mistakes, by identifying them, determining their source, and correcting them. Furthermore, people learn more from their own mistakes than from the successes of others.[*]

Russell L. Ackoff
Professor Emeritus of Management
Wharton School of the University of Pennsylvania

VALE CLYDACH REFINERY CASE STUDY

You have to get buy-in! If you don't get buy-in, you have to go back around and do it again. If you don't get buy-in from everybody, they just won't carry the task out as designed.

Damian Boucher
Manager, Clydach Nickel Refinery

Lean, or continuous improvement, is a lot more than just tools.

Fiona Buttrey
Technical Manager, Clydach Nickel Refinery

People are now empowered quite across the plant, not just in operations, but in service areas too. At the end of the day, it's a win-win for the business and the employees.

Phil Hayman
Continuous Improvement Manager, Clydach Nickel Refinery

Bob Fussel, superintendent of the process gas and ethanol plant at the Clydach Nickel Refinery, arrived at work expecting a normal day. Unfortunately, just before lunchtime, the power supply at the plant tripped, shutting down the entire refinery next door. This unanticipated, yet unfortunately common, occurrence was bad news for the refinery. Certainly, there were nickel pellets and other products in the middle of the refining process. Without power, those pellets would soon cool and harden. Wasted inventory was an unfortunate side effect of these unplanned downtimes. Bob thought to himself, "There has to be some way to stop the power from tripping unexpectedly." Unfortunately, there was no time to investigate the root cause of the power outage. He needed

---

[*] Ackoff, R. L. Thinking About the Future and Globalization. In2:InThinking Network, Inc. January 2006. Available at http://www.in2in.org/newsletter/articles/newsletter_2006_01a.htm.

to work with his team to get power restored to the refinery as quickly as possible in order to minimize downtime. When the power tripped at the gas plant, it could shut down the refinery's production plant for anywhere from 1 to 10 hours.

Thanks to recent continuous improvement initiatives at the Clydach Refinery, Bob and his team were able to find the root cause of the power outages. Using Lean tools such as the A3, he worked with the electrical department to solve the voltage and power trip problems. It turned out that when a low-voltage dip occurred, it took out one of the fans. That particular fan was linked to the trip system, so when it went off-line, it took the entire production plant off-line as well. They discovered in the starter panel a component called a phase protection, which was unnecessary for plant safety. When it was removed from the starter panel, it discontinued the connection between the fan and its ability to trip up the entire system. Since that problem was fixed, the factory has not had an unplanned power trip for 18 months. This was a relatively simple fix that brought a dramatic benefit. For 40 years those power trips had been occurring, but no one ever tried to get to the bottom of it. When they finally took the time, and did a bit of deep investigative work, it led to a fantastic benefit. This fix saved a lot of money and led to increased production levels as the production line wasn't stopping for unnecessary reasons.

History: What's the Story?

The Clydach Refinery is located near Swansea in South Wales, the United Kingdom. It is one of Europe's largest nickel refineries, producing high-purity nickel pellet and powder products for specialist applications such as high nickel alloys, batteries, nickel plating, and automotive components.

Ludwig Mond, a German-born chemist, established the refinery in 1902. He pioneered a unique and extremely complex process of refining nickel. The process takes nickel oxide, transforms it into nickel carbonyl gas, and then decomposes it into the pure nickel metal through heating. This process was so successful that the refinery would process imported nickel ore, which came from Canada. The refinery played a key role in Britain's success during World War I, because the refined nickel was combined with steel to create durable defensive armaments and projectiles. To this day, the Clydach factory uses the same process as invented by Ludwig Mond, virtually unchanged since the refinery was opened.

The Clydach Refinery has changed hands multiple times throughout its history. The latest acquisition took place in 2006, when the refinery's parent company, Canadian-based Inco, was acquired by Vale, a Brazilian mining company. The acquisition was inspired by Vale's desire to increase market share in the competitive mining market, and the acquisition of Inco gave Vale an entry into the European market. According to the most recent data (2011), Vale is the second biggest mining company in the world and the second biggest nickel producer. The mining and refining industry is quite competitive. The Clydach Refinery competes against other large refining companies such as MMC Norilsk Nickel based in Russia, the Jinchuan Group based in China, and Xstrata, based in Switzerland.

Currently, the refinery employs around 200 people and produces around 40,000 metric tons of nickel products per year. It supplies products to over 280 customers in over 30 countries in Europe, Asia, and the United States.

The Clydach Refinery remains successful today, over 100 years after its conception, due to the nature of its high-quality product and the innovative strategies that have been employed at the plant to keep production high. One such innovative strategy was the successful implementation of Lean principles, beginning in 2008. In fact, Lean principles were introduced at the refinery as early as 1995, but it wasn't until 2008 that the initiative really took off. After discovering the "Lean Iceberg Model" in 2009, the refinery was able to successfully integrate Lean thinking across all levels of the business and into the company culture.

Owing to the implementation of these Lean principles, the refinery has developed a vision that both their employees and clients can believe in. Currently, the company vision is to be the "refinery of choice" for their customers. At the time the interviews took place, the goal of the factory was 100 million pounds of annual nickel production. This ambitious goal is a metric that has never previously been achieved, but everyone at the refinery believes it is a possibility by using continuous improvement principles.

What Was the Need for Change?

After the 2006 acquisition by Vale, the Clydach Refinery needed to make changes in order to secure the long-term viability of the site. The refinery was facing many issues, including rising energy costs, an aging work-force, and deteriorating equipment. In addition, the refinery needed to reduce the headcount of their workforce. To stay profitable and relevant in their competitive industry, continuous improvement was introduced. Mike Cox, general manager of the refinery, stated, "I was convinced that continuous improvement could make a difference to our performance at Clydach."

There were many issues at the Clydach Refinery prior to 2008. As was previously mentioned, Lean principles had been introduced at the company as early as 1995, when they started documenting the responsibilities of each job. Various initiatives were implemented in the years that followed, but nothing stuck permanently. When company management decided in 2008 to apply Lean principles to all aspects of the refinery, they were facing an uphill battle. Hadn't Lean already been tested and failed before? Why should this time be different?

There were specific issues in many parts of the refinery that needed to be addressed as part of the continuous improvement journey. A sample of these issues will be highlighted here.

First, employee morale was low at the refinery. Historically, there was a top-down culture in the organization, as opposed to an open environment where employees felt empowered and enabled to contribute. There was the perception that a "chosen few" in the company made decisions, and everyone else just followed orders. In employees' eyes, there was no benefit for suggesting the company change and improve a process. Nobody wanted to try anything new because the fear of failure was too great.

Second, very few of the processes were standardized. There are a total of five shifts at the refinery, with multiple employees on each shift. Without standardized procedures to guide employee actions, each task could be completed in a minimum of five different ways. This was most apparent with the critical procedures in the pellet plant. Each of the operators on the five different shifts had a unique way of carrying out their task. Sometimes, they would find out that the task wasn't carried out correctly. This impacted production, quality, and sometimes the environment.

Continuous improvement principles were applied across the entire refinery, not just in the actual manufacturing plant. This meant that other business functions, such as HR and the environmental team, were required to implement Lean principles. This was a challenge from the outset for the environmental team because they functioned as a service department. How could they apply continuous improvement techniques to lessen the amount and types of waste the refinery produced?

Enacted Solutions and the Results

Although Lean thinking and Lean principles had been introduced at the refinery many years before, it wasn't until 2008 and 2009 that Lean finally stuck around for good in all areas. What was different about it this time? The individuals in charge of continuous improvement discovered the Lean Iceberg Model. This tool helped everyone at the company change the way they saw Lean. Previously, employees viewed Lean as a one-time event, called kaizen events in Lean terms. Kaizen events are quick, focused events, where employees focus on and improve a process. The life cycle for these events is very short. Once the kaizen event is done, the Lean thinking stops. With the Iceberg Model, Lean principles are incorporated into all facets of the company, including company strategy, leadership, and behavior. Lean thinking becomes an integral part of the company culture, not just an occasional business practice (see Figure 10.6).

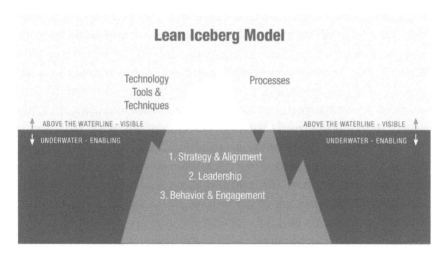

FIGURE 10.6
Lean Iceberg Model.

In order for Lean to finally succeed, Vale's culture at the refinery had to change. The historical top-down culture was not a conducive environment for Lean principles to take hold. Mike Cox, general manager of the refinery at the time, stated the following:

> *The big change in culture has been to take it from almost a 'fear of failure' to really empower people to take on responsibility and to not be afraid of the kind of repercussions of mistakes, because I think the only way you learn is through making mistakes.*

Employees needed to feel empowered and capable of contributing toward change. By allowing them to take risks, make suggestions, and even fail occasionally, employees quickly learned that the company wanted them to contribute, not to follow blindly. Mike also emphasized that in the last five to seven years getting everyone to feel they can contribute to the organization has been key. There isn't a group of "the chosen few" who make key decisions. It's about making sure everyone is included and taking his or her contributions into account.

The environmental team showed how much improvement a team can make when team members are engaged in Lean thinking and continuous improvement. As continuous improvement procedures began to be implemented across the entire refinery, some operators were unsure how to incorporate Lean principles into their work teams. The environmental team had a rough start because their team leader tried at first to force the department's hand. He made the mistake of telling his team members exactly how they were going to implement Lean. Once he realized his game plan was being met with heavy resistance, he began to increase the level of communication with his employees, explaining why the continuous improvement initiative was important. By engaging everyone around the site and using their suggestions, the environmental team began to have success. Since 2013, they've employed various Lean techniques, including KPIs around waste management. Another technique used was visual management to visually show the team's goals and progress. The team also used visual management to eliminate ambiguity in waste, which made the waste management process easier in all areas. The team's goal is to continuously find new innovations so that "Lean and Green" can cohabitate. One of the biggest successes they have in this department was reducing the waste management budget by 50%. As this team looks to the future, they have high hopes to continue implementing Lean in new ways. At the time the interviews took place, they had 83 current projects to reduce emissions through air and water. They want to be ahead of the game in terms of European Legislation regulatory requirements.

Another way that managers better engaged employees in continuous improvement and Lean thinking was through training. As managers initially sought to implement continuous improvement, they employed continuous improvement facilitators across all five shifts of the company. While these facilitators were able to implement Lean thinking to some extent, they were not getting the results management was looking for. Management and the facilitators also realized that the people on the actual shift created their own best practices, shared information, and created standardized work, linking continuous improvement to competence. Management decided to train more employees to implement continuous improvement, eventually training approximately one-third of the workforce in continuous improvement techniques. This created a more self-sustaining workforce and increased Lean thinking throughout the company, as frontline employees, being experts in their own area, were empowered to make changes immediately and improve efficiency. Management also saw an increase in teamwork and reliability in the company as a whole.

One of the greatest continuous improvement projects was the standardization of work procedures and the minimization of variations. The standard work project began in the pellet plant and eventually spread out across the organization. This continuous improvement project began by utilizing a value stream map. First, each of the operators was observed carrying out their job responsibilities. Everything they were doing was written down, including the time it took them to do it. It was discovered that five different methods existed to accomplish a single task. When every single task the operator carried out throughout that area and all different areas of the refinery were recorded, there were a total of 150 tasks. Each of those tasks was risk assessed, and only 40 were identified as critical tasks. Those 40 tasks formed the basis of the "ONE" ideal method to carry out that job responsibility. That value stream mapping process became extremely critical to the company because it streamlined those tasks, eliminated waste, and saved time and money. The project initially was met with resistance by operators because they didn't understand why they were being observed and analyzed. Once they understood that the purpose of the observation was to discover how to get the task carried out in the safest, most efficient way, they jumped on board. It took 18 months to complete the value stream map and critical task assessment, but that time was well worth the investment. A post-analysis was conducted and the data showed that the process was much more stable than before.

Value Stream Mapping & Theory of Constraints

Theory of constraints (TOC) works on the concept that "A chain is no stronger than its weakest link".

Looking at constraints in the process cost reduction opportunities, plant performance and reliability projects

Three necessary conditions

1. Define the system
2. State the purpose
3. Decide how to measure it

Five Focusing Steps

1. Identify the system's constraint(s).
2. Decide how to explore the system's constraint(s).
3. Subordinate everything else to the above decisions.
4. Elevate the system's constraint.
5. When a constraint is broken, return to step one. Do not allow inertia to become the system's constraint.

As a direct result of the success of the standardized work system in the pellet plant, a new system called Problem Follow Up (PFU) was created. The idea behind the PFU system was for employees to identify problems (by writing them down) and then coming up with a solution so that problem never happened again. As employees carried out the new standardized procedures and something went wrong, they identified it on a PFU slip, and implemented the change at the next opportunity. The PFU system was directly linked to the standardized work because it allowed all employees the opportunity to offer suggestions to continuous improvement throughout the plant. The system worked very well. In the first year, PFUs were linked to the operator's annual incentive plan (their bonus). Each operator was required to complete at least two PFUs per year. This helped incorporate Lean thinking and processes into the refinery's culture. As the PFU system gained popularity, it was eventually rolled out to all areas in the refinery, even those not using standardized work. This further increased the emphasis on Lean thinking and continuous improvement in the company culture.

In summary, the two biggest challenges that employees and management at Clydach faced in implementing continuous improvement and Lean thinking were company culture and standardized work processes. By changing employee engagement through training, listening to all employees and empowering employees to make decisions, employees became engaged and enthusiastic about Lean thinking. Then, by using value stream mapping to identify the ideal methods for work processes, standardized work processes were implemented throughout the refinery. This increased efficiency and minimized waste for each process. When implemented effectively, continuous improvement can spread rapidly across all levels of the organization, changing the way each employee sees their work. The Clydach Refinery has seen improvements throughout all facets of its business, in areas such as employee satisfaction, decreased down time, increased savings, and much more.

Discussion Questions

1. Why was Vale Clydach unsuccessful in implementing Lean principles on their first attempt?
2. What changed in 2008 to make them successful in implementing Lean principles?
3. Using this example, how best should managers implement Lean principles in other organizations?
4. Why is employee engagement necessary to implement Lean principles?

11

The Shingo Assessment Process

We are what we repeatedly do. Excellence, then, is not an act but a habit.[*]

Will Durrant
Author

ASSESSMENT

So far this book has given the reader tools to use for an internal assessment process. However, a clear and valid assessment can be achieved only by using an external, neutral assessment tool. This is true for a variety of reasons:

- Organizations don't know what they don't know—External experts can identify opportunities that were not visible to them because they are so integrated in the work.
- Companies aren't where they think they are—This can go in both directions. They may not be as bad in some areas as they think they are, but they're also not as good as they think they are in some other areas.
- An independent point-of-view is needed.
- A good assessment will raise a different set of questions; it will identify opportunities that insiders didn't even consider.
- An assessment should identify the "Big Gaps" that link to the priorities.
- Prioritization of the gaps—Use tools like the Impact/Effort matrix to identify quick wins and big wins.

Understanding there is a lot to learn from an external assessment, the next step is to realize there are a lot of assessment tools available. Some

---

[*] Durant, W. Aristotle and Greek Science. *The Story of Philosophy: The Lives and Opinions of the World's Greatest Philosophers*. New York, NY: Simon & Schuster, 1953.

Respect Every Individual

Support
We invest in everyone's development and encourage them to realize their potential.

	Strongly Disagree								Strongly Agree		The question isn't clear	I do not know the answer	This isn't relevant to my work
My department invests time and energy developing other's potential.	○	○	○	○	○	○	○	●	○	○	○	○	○
	○	○	○	○	○	○	○	○	○	○	○	○	○

Recognition
We honor contributions of every employee.

	Strongly Disagree								Strongly Agree		The question isn't clear	I do not know the answer	This isn't relevant to my work
	○	○	○	○	○	○	○	●	○	○	○	○	○
	○	○	○	○	○	○	●	○	○	○	○	○	○

FIGURE 11.1
Shingo Insight Survey Page.

are good in that they are principle focused. And some are not so good because they are tools focused. The Shingo Institute recommends the Shingo Insight assessment process, because it directly focuses on the *Shingo Guiding Principles* and gives the surveyed company a tool that will allow them to have insight to their strengths and shortcomings.

Figures 11.1 and 11.2 are examples from a Shingo Insight online assessment. Figure 11.1 shows an example of the survey instrument that is used,

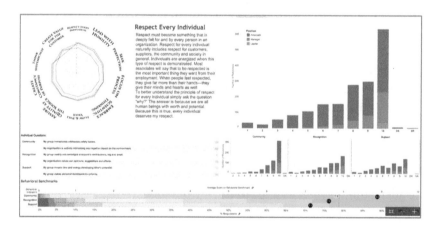

FIGURE 11.2
Shingo Insight Sample Report.

and Figure 11.2 shows an example of one page of the interactive, online report.

The benefits of using Shingo Insight are twofold. First, this internal assessment provides a reasonably quick, value-oriented, and easy path to establishing the baseline of the organizational culture, finding gaps to address, and understanding how an organization aligns to the guiding principles in the *Shingo Model*. Secondly, when combined with the feedback organizations receive from a team of Shingo examiners when they challenge for the Shingo Prize, these two vantage points, both internal and external, can complete the whole picture of where an organization stands in relation to Shingo Prize standards and how it can accelerate its journey of operational excellence.

For more details about the Shingo Insight survey process, please visit http://shingo.org/insight.

THE SHINGO PRIZE AS AN ASSESSMENT TOOL

Another form of assessment that is utilized by a large number of companies is the Shingo Prize assessment process. As learned in Chapter 2, the Shingo Prize has become the world's highest standard for enterprise excellence. As an effective way to benchmark progress toward enterprise excellence, organizations throughout the world may apply and challenge for this Prize. Recipients receiving this recognition fall into these three categories:

SHINGO BRONZE MEDALLION
SHINGO SILVER MEDALLION
SHINGO PRIZE

Most organizations do not wait until they believe they might qualify for the Shingo Prize to challenge. They challenge for the Prize so they can have a team of enterprise excellence experts, Shingo examiners, visit their organization and evaluate their culture and performance.

Some organizations do not intend to challenge for the Prize, but use the *Shingo Model* and the Prize assessment process to measure themselves as they work toward the highest standard of excellence in the world. They use these guidelines as a tool to direct them.

For more information about the Prize, and to see the hundreds of companies who have challenged for and received the Prize, see the Appendix. It's also useful to visit the Shingo Institute website to learn more about the Shingo Prize assessment criteria. Go to http://shingo.org/challengefortheprize, scroll down the page, and click the "Application Guidelines" tab to download the guidelines and assessment criteria.

DOES "A COUPLE OF TOOLS" WORK BETTER?

There are a few tools that are extremely helpful as one attempts to improve the behaviors and drive toward enterprise excellence. These tools are the impact/effort matrix and the A3 problem-solving worksheet.

Impact/Effort Matrix

As discussed earlier, there are several forms of assessment throughout this book including the Shingo Impact survey. Whatever assessment tool the reader uses, the result is that they will end up with a long list of gaps, and it becomes next to impossible to work on all the different gaps simultaneously. This means the reader will need to prioritize the gaps, identifying which will be the most beneficial and targeting those first. Sometimes that prioritization is obvious. For example, the cultural enabler priorities should always be on the top of everyone's list. There is a tool to assist with

Behavioral benchmarks will be subjectively prioritized by the team to identify the impact and effort priorities.

The behavioral benchmarks in the upper left-hand quadrant will receive the most immediate focus, followed by the upper right-hand and lower left-hand quadrants.

At this stage the focus in the Impact/Effort matrix is more qualitative than quantitative. Objective methodology still needs to be applied.

The **Impact/Effort Matrix** is a lean tool used to prioritize projects and activities.

FIGURE 11.3
Impact/Effort Matrix.

the prioritization process that the Shingo Institute often uses. It is known as the Impact/Effort Matrix (see Figure 11.3).

For example, if one goes through the Shingo process, they will end up with a list of Behavioral Benchmark gaps, each of which needs attention at some point in time. List those gaps and assign a number to each of them. Then draw the graphic as shown in Figure 11.3 with its four quadrants. The vertical dimension charts the impact of the gap and the horizontal charts the effort. There is no scale assigned to the chart, but if the reader feels a scale is appropriate, one could easily be assigned. All the gaps are considered relative to each other for graphing purposes.

The number of each gap is posted on the chart relative to its impact and effort to the other gaps that have been identified. Soon the chart is filled with numbers. One caution is that this is a relative mapping (how each gap would map out relative to any of the other gaps) and if the reader gets wrapped up in the data, like estimating an exact cost or timeline, they'll get bogged down. That level of accuracy is not only necessary, but it also negatively affects the process.

Once all the gaps have been subjectively prioritized on the chart, start to identify what to work on first. Begin in the upper left-hand corner and work diagonally down toward the lower right-hand corner. Start with the first project and work down until the resources run out. Then, as a project is successfully completed, add new projects to the list. One danger, however, is that as projects are tackled, the reader will identify new projects and the number of projects will increase rather than decrease, which means the reader will identify more and more opportunities for improvement as they progress. That's a good thing. One will rarely get to the end of the project list because if they did they would be perfect, and the Shingo Institute has yet to see that happen to anyone.

A3 Problem-Solving Worksheet

Now that the reader has successfully identified the focus of their efforts, document the project by identifying a champion for the project (someone who is willing to pay the bill) and forming a team around the project. The qualitative methodology used when constructing the impact/effort matrix should now be replaced with more quantitative values. The boss will want to know how much this will cost and how long it will take. He or she may also want the reader to spell out the benefits of doing this exercise. The tool for this is the A3 (see Figure 11.4).

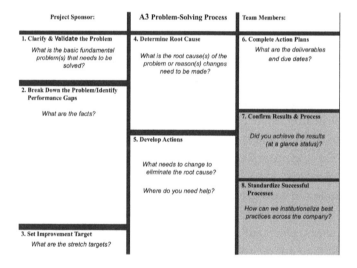

FIGURE 11.4
A3 Problem-Solving Worksheet Example.

The purpose of the A3 is to show everything on one sheet of paper. As shown in the figure, the A3 flows through the project. It contains the following project steps:

1. Clarify and Validate the Problem (or Opportunity): What is the basic fundamental change one is trying to make with this project? Give a description of the project in a couple sentences.

2. Break Down the Problem and Identify the Performance Gaps: This is where the numbers come in. What are the detailed facts that demonstrate the problem? Sometimes, when one works on a principle, numbers don't apply because the goal is to change the culture, and how does one quantify that? But even then, this is where one explains and justifies what it is they hope to change.

3. Set Improvement Targets: What is the goal the reader attempts to achieve by going through this change?

4. Determine the Root Cause: If it's a problem one is trying to fix, what is the root cause of the problem? If it is an improvement one is focused on, what is it that is not working as well as it should?

5. Develop Actions: Create a step-by-step list of activities that need to be completed in order to implement this change. This is also the point where the reader specifies what resources are needed in order to accomplish the change, whether its money, or people, or time. This is where it is indicated.

6. Complete Action Plans: Take the list from step 5 and put a timeline on it including a specific list of all the deliverables one hopes to accomplish. At this point, all the steps necessary to get project approval are completed.

7. Confirm Results and Process: Post progress to the timeline, the deliverables listed in item 6, and the target values listed in item 3.

8. Standardize Successful Process: Turn the new process into standard work, and share the learning with other relevant elements throughout the enterprise. Institutionalize this best practice across the company.

With these two tools in place the reader is ready to press forward with implementation of the changes, which will drive ever closer to the goal of enterprise excellence.

Defects are results - not causes. Therefore, checking once defects have occurred can never eliminate defects entirely.[*]

Shigeo Shingo

---

[*] Shingo, S. *The Shingo Production Management System: Improving Process Functions.* Portland, OR: Productivity Press, 1988.

Section IV

Wrap Up

12

Bringing It All Together

If you want to live a happy life, tie it to a goal.[*]

Albert Einstein

SHINGO PRIZE RECIPIENT

The Shingo Institute concludes this book with a set of employee testimonials from a 2016 Shingo Prize recipient, Rexam Beverage Can Americas S. A. de C.V (now Ball Corporation) in Queretaro, Mexico. Please consider these comments relative to their Shingo experience.

> *Rexam is a global manufacturer of aluminum containers with 51 plants and a presence in over 23 countries. Our plant is located in Queretaro, Mexico, and we are partners of some of the most successful brands of beverages; our vision is "to be the best beverage can maker in the world" and our culture is based on five values (teamwork, safety, continuous improvement, recognition, and trust).*

Juan Jose Sanchez
General Manager

> *At Rexam, we direct our efforts to achieve sustainable continuous improvement in our processes, products, and services; this way we deliver value to our customers. We develop operational excellence based on the fact that we, the associates, are the most important part of our organization. The #1 priority*

---

[*] Einstein, A. Albert Einstein Quotes. Quotes.net. STANDS4 LLC, 2017. Available at http://www.quotes.net/quote/9407.

is our safety. We support our strategies through the balanced scorecard in which we define the strategic business objectives.

Maria Elena Suarez
Plant Manager

At Rexam, our associates are most important; this being a business strategy to construct a winning organization. We focus our efforts to work in a safe environment with recognition, trust, and respect for the individual. Also, to develop the personnel to obtain the best results to offer customers the best aluminum containers.

Mario Dominguez
HR Manager

At Rexam, we are interested in being a partner of our customers and antici-pate their needs. We maintain a close relationship with them, and we make sure we always find new ways to add value to our products. To achieve this, we offer to our customers innovation, specialized technical assistance, immediate emergency response, and record time in development of new products. A manner in which we have achieved alignment to create value and deliver results has been through our communication system. One of the pillars has been group meetings where we hear, we train, we recognize and work together to seek continuous improvement in everything that we do!

Hector Marin
Marketing and Customer Service Manager

At Rexam, continuous improvement is one of our values: seeking perfection, listening to the ideas of our partners, and applying the talent of people is vital for sustainable results. Working with these principles generates commitment and ideal behaviors. All members receive training in systems such as TPM and 5S to create autonomous teams.

The adoption of the Shingo Model has driven our systems and put us on the path of operational excellence.

David Alvarado
Assistant Plant Manager

Working with systems and tools based on principles and ideal behaviors has given us guidelines toward operational excellence. At Rexam, the respect for the individual and the care of the environment are fundamental.

Jaime Verdi
EHS Manager

Applying Lean tools like SMED, kaizen teams, VSM, Six Sigma, among others, has allowed us to focus efforts as well as to use and develop the talents of our associates. Continuous improvement is not only a methodology, it is a work form that allows us to identify opportunity areas and ensure business results to be a world-class company.

Carlos Macias
Production Manager

Assuring quality at the source guarantees that our customer requirements are effectively transformed into process variables. These are monitored through inspections carried out by our associates to ensure that the product we manufacture meets the defined specifications. The quality control plans, procedures, and operating instructions are part of our work standards to control the variability of our processes.

Urbano Fernandez
Quality Manager

Every day we strive, through enterprise alignment, to develop effective meetings to follow-up on issues that generate value to customers and our organization. We listen assertively to the contributions of all, making daily and strategic decisions so as to focus our efforts in the right direction. At Rexam, we passionately enjoy doing our job.

Francisco Vega
Logistics Manager

The best ideas always come from those who do the work every day, so at Rexam, we always listen carefully to the concerns of our team. That's leading with humility.

Melesio Lopez
Warehouse and Shipping Manager

At Rexam, we work to build a winning organization.

Manuel Garcia
Forklift Operator

I am happy to work in this company because to Rexam I am the most important.

Elizabeth Perez
Quality Inspector

Teamwork has been the key to continuous improvement.

Luis Valdez
Back-End Mechanic

Company Message

The Shingo Model *has been the key for our organizational development and that's why we confirm our vision, "to be the best beverage can maker in the world."*

RECAP

This chapter summarizes what was covered in this book. But first reflect back to the Einstein challenge presented in the book's introduction. If the reader doesn't remember it, go back to the beginning of the book and review it before looking at the answer, which will appear later in this chapter. Now for a review of what was discussed:

- The book started with the question, "What is enterprise excellence?"
- There was a discussion of long-term sustainability and whether continuous improvement tools were sufficient. Or is there something more one needs in order to sustain the achieved improvements?
- This led to a discussion about the importance of results, which included KPIs versus KBIs. Behaviors and forward-looking KBIs are more important for long-term sustainability than KPIs, which are backward focused.
- The next conversation was about behaviors and the role they play in creating culture. It was stressed that cultural shift is critical to sustainable continuous improvements.
- Quite a bit of time was spent on culture and its importance, both good and bad, how it's created, and who is responsible for it.
- The *Three Insights of Enterprise Excellence* were addressed, along with their relevance to cultural change.
- A discussion of the *Shingo Model* and its components included tools, systems, results, guiding principles, and of course culture.
- The *Shingo Guiding Principles* were then broken down into four dimensions with a total of 10 principles in all. Each of these principles

was reviewed in detail with a discussion of what they mean to the enterprise.

- From there, the reader learned how to assess the current condition of their own enterprise. Where are the gaps and how does one identify them?
- There was also a discussion about the importance of the "go and observe" walks within an organization and how this should be used to identify behavioral gaps.
- Ways to prioritize the gaps and move forward with the continuous improvement process, which is designed to eventually drive the culture of an organization toward enterprise excellence, were also shared.

In summary, the reader learned that

- The goal is to transform an organization's culture to one that supports sustainable continuous improvements and creates an enterprise excellence culture.
- The purpose of the 10 *Shingo Guiding Principles* is to inform ideal behaviors.
- Systems and purpose drive behaviors.
- Changing systems changes behaviors, which in turn defines the culture of an organization.

Now to the answer of Einstein's problem in the introduction of this book. It is Einstein's nationality—German.

In summary, to become excellent, one needs to pursue excellence. No one wants to imitate what someone else is doing, including Toyota. One needs to find the culture within for what they want the enterprise to become. And then they need to go get it!

The Toyota style is not to create results by working hard. It is a system that says there is no limit to people's creativity. People don't go to Toyota to 'work,' they go to Toyota to 'think.'[*]

Taiichi Ohno
Former CEO, Toyota

---

* Yorke, C., Bodek, N. *All You Gotta Do Is Ask*. Vancouver, WA: PCS Press, 2005.

Appendix

SHINGO AWARD RECIPIENTS

Year	Level	Company Name	City	State	Country	Industry
2017	Silver	Visteon Electronics Tunisia—Bir El Bey Plant	Bir El Bey	Tunis	Tunisia	Manufacturing
2017	Silver	MassMutual, CFO and MMUS Operations	Springfield	Massachusetts	USA	Financial
2017	Shingo Prize	Thermo Fisher Scientific, Vilnius	Vilnius		Lithuania	Manufacturing
2017	Bronze	Cardinal Health, Quiroproductos de Cuauhtemoc, S. de R.L. de C.V.	Cuauhtemoc	Chihuahua	Mexico	Medical
2017	Shingo Prize	Ball Beverage Packaging Europe, Naro Fominsk Ends	Naro Fominsk	Moscow Oblast	Russia	Consumer Goods
2017	Bronze	Land Apparel S.A.	Puerto Cortés	Puerto Cortés	Honduras	Consumer Goods
2017	Bronze	Letterkenny Army Depot, PATRIOT Launcher New Build Program	Chambersburg	Pennsylvania	USA	Military
2016	Silver	Hospira Limited Ltd, a Pfizer Company	Haina	San Cristobal	Dominican Republic	Pharmaceutical
2016	Silver	Meda Rottapharm Ltd—a Mylan company	Dublin	Dublin	Ireland	Pharmaceutical
2016	Shingo Prize	Boston Scientific Cork	Cork	Cork	Ireland	Pharmaceutical
2016	Shingo Prize	Rexam Beverage Can Americas Querétaro	Querétaro	Querétaro	México	Consumer Goods
2015	Silver	Commonwealth Bank of Australia, Collections & Customer Solutions	Sydney	New South Wales	Australia	Financial
2015	Bronze	Boston Scientific, Coyol	El Coyol	Alajuala	Costa Rica	Medical

(*Continued*)

Year	Level	Company Name	City	State	Country	Industry
2015	Shingo Prize	Envases Universales Rexam de Centro America, S.A.	Amatitlan	Guatemala	Guatemala	Consumer Goods
2015	Bronze	Lake Region Medical	New Ross	Wexford	Ireland	Medical
2015	Shingo Prize	Abbott Diagnostics Longford	Longford	Longford	Ireland	Medical
2015	Bronze	Carestream Health, Yokneam	Yokneam	Yokneam	Israel	Medical
2014	Bronze	Vistaprint Deer Park Australia	Derrimut	Victoria	Australia	Printing
2014	Silver	Rexam Beverage Can, Enzesfeld	Enzesfeld	Vienna	Austria	Consumer Goods
2014	Silver	Rexam Beverage Can South America, Jacarei	Jacarei	Sao Paulo	Brazil	Consumer Goods
2014	Bronze	Rexam Beverage Can South America, Rio de Janeiro	Rio de Janeiro	Rio de Janeiro	Brazil	Consumer Goods
2014	Bronze	Corporation Steris Canada	Quebec	Quebec	Canada	Medical
2014	Bronze	Autoliv (China) Steering Inflator Co., Ltd.	Shanghai	Shanghai	China	Automotive
2014	Silver	Rexam Healthcare, Neuenburg	Neuenburg am Rhein	Baden-Württemberg	Germany	Medical
2014	Shingo Prize	Abbott Vascular	Clonmel	Tipperary	Ireland	Medical
2014	Shingo Prize	DePuy Synthes	Cork	Cork	Ireland	Medical
2014	Bronze	Lundbeck Pharmaceuticals Italy SPA	Padova	Padua	Italy	Pharmaceutical
2014	Silver	Plasticos y Materias Primas (PyMPSA)	Guadalajara	Jalisco	Mexico	Pharmaceutical
2014	Shingo Prize	News UK-Newsprinters Ltd	Holytown	Motherwell	UK	Printing
2014	Silver	Vale Europe Ltd Clydach Refinery	Clydach	Swansea	UK	Chemical
2014	Silver	Boston Scientific, Maple Grove Operations	Maple Grove	Minnesota	USA	Medical
2014	Shingo Prize	Barnes Group Inc. Acting Through Its Barnes Aerospace OEM Strategic Business Unit	Ogden	Utah	USA	Aviation & Aerospace
2013	Silver	Rexam Beverage Can South America, Manaus Ends	Manaus	Amazonas	Brazil	Consumer Goods

(Continued)

Year	Level	Company Name	City	State	Country	Industry
2013	Bronze	Rexam Beverage Can South America, Cuiaba Cans	Cuiaba	Mato Grosso	Brazil	Consumer Goods
2013	Silver	Rexam do Brasil Ltda Extrema Can Plant	Extrema	Minas Gerais	Brazil	Consumer Goods
2013	Silver	Visteon Climate Systems India Ltd	Bhiwadi, Alwar	Rajasthan	India	Automotive
2013	Silver	Visteon Electronica Mexico—Saucito Plant	Chihuahua	Chihuahua	Mexico	Automotive
2013	Silver	MEI Queretaro	El Marques	Queretaro	Mexico	Computer & Electronic
2013	Bronze	Starkey de Mexico S.A. de C.V.	Matamoros	Tamaulipas	Mexico	Medical
2013	Silver	Pentair Water Pool and Spa	Moorpark	California	USA	Consumer Goods
2013	Bronze	Regeneron Pharmaceuticals Inc (IOPS)	Rensselaer	New York	USA	Medical
2013	Bronze	Letterkenny Army Depot, Force Provider	Chambersburg	Pennsylvania	USA	Military
2012	Shingo Prize	Rexam Beverage Can, Aguas Claras Cans	Aguas Claras	Rio Grande do Sul	Brazil	Consumer Goods
2012	Bronze	Remy Automotive Brasil Ltda.	Brusque	Santa Catarina	Brazil	Automotive
2012	Bronze	Lake Region Medical Limited	New Ross	Co. Wexford	Ireland	Medical
2012	Shingo Prize	Ethicon Inc	Juarez	Chihuahua	Mexico	Medical
2012	Silver	Visteon Electronica Mexico—Carolinas Plant	Chihuahua	Chihuahua	Mexico	Automotive
2012	Bronze	Johnson Controls Lerma Plant	Lerma	Mexico	Mexico	Automotive
2012	Silver	Pentair Technical Products	Reynosa	Tamaulipas	Mexico	Automotive
2012	Bronze	State Farm Life Insurance Company, Operations Center	Bloomington	Illinois	USA	Financial
2012	Silver	Tobyhanna Army Depot, COMSEC	Tobyhanna	Pennsylvania	USA	Military
2011	Silver	Rexam Beverage Can South America-Recife Ends	Cabo Sto Agostinho	Cabo Sto Agostinho	Brazil	Consumer Goods
2011	Bronze	Rexam Plastic Packaging do Brasil	Jundiai	Sao Paulo	Brazil	Consumer Goods
2011	Shingo Prize	Goodyear do Brasil Produtos de Borracha Ltda	Americana	Sao Paulo	Brazil	Automotive

(Continued)

Year	Level	Company Name	City	State	Country	Industry
2011	Silver	Autoliv (China) Steering Wheel Co., Ltd.	Shanghai	Shanghai	China	Automotive
2011	Silver	Lundbeck, Supply Operation & Engineering (Valby and Lumsas site)	Valby	Copenhagen	Denmark	Medical
2011	Silver	dj Orthopedics de Mexico S.A.de C.V.	Tijuana	Baja California	Mexico	Medical
2011	Silver	Remy Components, S. de R.L. de C.V.	San Luis Potosi	San Luis Potosi	Mexico	Automotive
2011	Bronze	Leyland Trucks Ltd	Leyland	Lancashire	UK	Automotive
2011	Bronze	Denver Health, Community Health Services	Denver	Colorado	USA	Healthcare
2011	Bronze	U.S. Army Armament Research, Development & Engineering Center	Picatinny Arsenal	New Jersey	USA	Military
2011	Bronze	Letterkenny Army Depot, Aviation Ground Power Unit	Chambersburg	Pennsylvania	USA	Military
2011	Silver	Tobyhanna Army Depot (AN/MST -T1(V)), MiniMutes	Tobyhanna	Pennsylvania	USA	Military
2011	Shingo Prize	U.S. Synthetic	Orem	Utah	USA	Heavy Equipment
2011	Silver	Barnes Group Inc. Acting Through Its Barnes Aerospace OEM Strategic Business Unit	Ogden	Utah	USA	Defense Contractor
2010	Silver	Hi-Tech Gears Ltd.	Manesar, Gurgaon	Haryana	India	Automotive
2010	Silver	Autoliv Steering Wheels Mexico AQW S. de R.L. de C.V.	El Marques	Queretaro	Mexico	Automotive
2010	Silver	Goodyear Tire & Rubber	Lawton	Oklahoma	USA	Automotive
2010	Bronze	Letterkenny Army Depot, Patriot Missile	Chambersburg	Pennsylvania	USA	Military
2010	Bronze	Tobyhanna Army Depot, AIM-9M Sidewinder Missile	Tobyhanna	Pennsylvania	USA	Military
2010	Shingo Prize	Lycoming Engines	Williamsport	Pennsylvania	USA	Aviation & Aerospace
2010	Shingo Prize	John Deere, Power Products	Greeneville	Tennessee	USA	Consumer Goods

(Continued)

Year	Level	Company Name	City	State	Country	Industry
2009	Bronze	Visteon Interamerican Plant	Apodaca	N.L.	Mexico	Automotive
2009	Shingo Prize	Guanajuato Manufacturing Complex North Plant	Silao	Silao	Mexico	Automotive
2009	Shingo Prize	Gulfstream Aerospace, Interiores Aéreos S.A. De C.V.	Mexicali	Mexicali	Mexico	Aviation & Aerospace
2009	Silver	Valeo Sylvania Iluminacion	Queretaro	Queretaro	Mexico	Automotive
2009	Bronze	Ultraframe UK Ltd.	Clitheroe	Lancashire	UK	Building Material
2009	Bronze	BAE Systems–Samlesbury	Blackburn	Lancashire	UK	Defense Contractor
2009	Bronze	Aviation Center Logistics Command and Army Fleet Support, Lowe Army Heliport	Ft. Rucker	Alabama	USA	Military
2009	Silver	Lockheed Martin Missiles and Fire Control, Camden Operations	East Camden	Arkansas	USA	Defense Contractor
2009	Bronze	Baxter Healthcare	Los Angeles	California	USA	Medical
2009	Silver	HID Global	North Haven	Connecticut	USA	Computer & Electronic
2009	Bronze	Fleet Readiness Center Southeast, TSRS Shop	Jacksonville	Florida	USA	Military
2009	Shingo Prize	E-Z-GO	Augusta	Georgia	USA	Consumer Goods
2009	Silver	402D Electronics Maintenance Group, Warner Robins Air Logistics Center, Robins Air Force Base	Warner Robins	Georgia	USA	Military
2009	Silver	Carestream Health Inc., Rochester Finishing	Rochester	New York	USA	Medical
2009	Bronze	Red River Army Depot, Up-Armored High Mobility Multipurpose Wheeled Vehicle (UAH)/HEAT	Texarkana	Texas	USA	Military
2009	Shingo Prize	Autoliv Airbag Module Facility	Ogden	Utah	USA	Automotive
2009	Shingo Prize	Autoliv Inflator Facility	Brigham City	Utah	USA	Automotive

(*Continued*)

Year	Level	Company Name	City	State	Country	Industry
2009	Silver	EFI Electronics by Schneider Electric	Salt Lake City	Utah	USA	Building Material
2008	Shingo Prize	Baxter	Cartago	Cartago	Costa Rica	Medical
2008	Silver	DJ Orthopedics de Mexico S.A.de C.V.	Tijuana	Baja CA	Mexico	Medical
2008	Silver	Delphi Sistemas de Energia S.A.de C.V.	Torreon	Coahuila	Mexico	Automotive
2008	Silver	American Axle and Manufacturing de Mexico North Plant	Silao	Guanajuato	Mexico	Automotive
2008	Bronze	IACNA Mexico	Toluca	Toluca	Mexico	Automotive
2008	Shingo Prize	Denso Mexico S.A. de C.V. Guadalupe Plant	Guadalupe	Guadalupe	Mexico	Automotive
2008	Shingo Prize	KEMET Electronics	Matamoros	Matamoros	Mexico	Computer & Electronic
2008	Shingo Prize	KEMET Electronics	Victoria	Victoria	Mexico	Computer & Electronic
2008	Shingo Prize	Carestream Health, Inc	Guadalajara	Guadalajara	Mexico	Medical
2008	Shingo Prize	ZF Lemforder Corporation	Tuscaloosa	Alabama	USA	Automotive
2008	Bronze	Raytheon Missile Systems	Camden	Arkansas	USA	Military
2008	Silver	Lockheed Martin Missiles & Fire Control	East Camden	Arkansas	USA	Military
2008	Silver	Fleet Readiness Center Southwest, E2C2 Product Line	San Diego	California	USA	Military
2008	Bronze	Carestream Health, Converting Operations	Windsor	Colorado	USA	Medical
2008	Bronze	HID Global	North Haven	Connecticut	USA	Computer & Electronic
2008	Silver	Medtronic Surgical Technologies, Jacksonville Plant	Jacksonville	Florida	USA	Medical
2008	Silver	Fleet Readiness Center Southeast, F/A-18 Hornet Center Barrel Replacement Program	Jacksonville	Florida	USA	Military
2008	Silver	Lockheed Martin Missiles and Fire Control	Orlando	Florida	USA	Military
2008	Silver	Chrysler ITP II	Kokomo	Indiana	USA	Automotive

(Continued)

Year	Level	Company Name	City	State	Country	Industry
2008	Bronze	Callaway Golf Ball Operations	Chicopee	Massachusetts	USA	Consumer Goods
2008	Bronze	Extrusion Technology	Randolph	Massachusetts	USA	Food & Beverages
2008	Silver	Raytheon Integrated Defense Systems, Integrated Air Defense Center	Andover	Massachusetts	USA	Computer & Electronic
2008	Bronze	Delphi Steering, Plant 14	Saginaw	Michigan	USA	Automotive
2008	Shingo Prize	Metalworks/Great Openings	Ludington	Michigan	USA	Consumer Goods
2008	Silver	Global Engine Manufacturing Alliance (GEMA), North Plant	Dundee	Michigan	USA	Automotive
2008	Silver	Micron Manufacturing Company	Grand Rapids	Michigan	USA	Machining
2008	Silver	Volvo Construction Equipment North America, Asheville Plant	Skyland	North Carolina	USA	Automotive
2008	Silver	Fleet Readiness Center East, AV-8B Aircraft Production Program	Cherry Point	North Carolina	USA	Military
2008	Shingo Prize	Sandia National Laboratories, Responsive Neutron Generator Product Development Center	Albuquerque	New Mexico	USA	Military
2008	Bronze	Tinker AFB 565th AMXS/MXPA, B-1B Bomber	Tinker AFB	Oklahoma	USA	Military
2008	Bronze	Letterkenny Army Depot, Biological Integrated Detection System (BIDS)	Chambersburg	Pennsylvania	USA	Military
2008	Bronze	Tobyhanna Army Depot, AN/ASM-189 Maintenance Electronic Shop Van	Tobyhanna	Pennsylvania	USA	Military
2008	Bronze	Tobyhanna Army Depot, AN/TYQ-23 Tactical Air Operations Module	Tobyhanna	Pennsylvania	USA	Military
2008	Silver	Lycoming Engines	Williamsport	Pennsylvania	USA	Aviation & Aerospace

(*Continued*)

Year	Level	Company Name	City	State	Country	Industry
2008	Silver	Roche Carolina Inc.	Florence	South Carolina	USA	Healthcare
2008	Bronze	Comanche Peak Nuclear Power Plant	Glen Rose	Texas	USA	Chemical
2008	Bronze	Red River Army Depot, Patriot Missile	Texarkana	Texas	USA	Military
2008	Bronze	Red River Army Depot, Trailer	Texarkana	Texas	USA	Military
2008	Silver	Red River Army Depot, Heavy Expanded Mobility Tactical Truck (HEMTT)	Texarkana	Texas	USA	Military
2008	Silver	Luminant, Martin Lake Complex	Martin Lake	Texas	USA	Other
2008	Shingo Prize	Autoliv, Inflator Facility	Brigham City	Utah	USA	Automotive
2007	Shingo Prize	Autoliv Querétaro CMX Facility	Queretaro	Queretaro	Mexico	Automotive
2007	Shingo Prize	Delphi Packard Electrical/Electronic Architecture's Cableados Fresnillo 1	Fresnillo	Fresnillo	Mexico	Automotive
2007	Shingo Prize	Delphi Packard Electrical/Electronic Architecture's Chihuahua 1	Chihuahua	Chihuahua	Mexico	Automotive
2007	Shingo Prize	Takata Seat Weight Sensor, Equipo Automotriz Americana	Monterrey	Monterrey	Mexico	Automotive
2007	Shingo Prize	Solectron Manufactura de Mexico	Guadalajara	Guadalajara	Mexico	Communication
2007	Shingo Prize	Cordis de Mexico	Chihuahua	Chihuahua	Mexico	Healthcare
2007	Shingo Prize	Baxter S.A. de C.V., Cuernavaca, Mexico Plant	Jiutepec Morelos	Jiutepec Morelos	Mexico	Medical
2007	Bronze	Anniston Army Depot, Turbine Value Stream	Anniston	Alabama	USA	Military
2007	Bronze	Aviation Center Logistics Command and Army Fleet Support, TH-67 C20J Engine Project	Ft. Rucker	Alabama	USA	Military
2007	Finalist	ZF Lemforder Corporation	Tuscaloosa	Alabama	USA	Automotive

(*Continued*)

Year	Level	Company Name	City	State	Country	Industry
2007	Silver	Anniston Army Depot, Vehicle Value Stream, Tracked Systems Division	Anniston	Alabama	USA	Military
2007	Shingo Prize	The HON Company, LA South Gate Plant	South Gate	California	USA	Other
2007	Silver	Naval Sea System Command, Port Hueneme Division (PHD), RAM Value Stream	Port Hueneme	California	USA	Military
2007	Bronze	Fleet Readiness Center Southeast (FRCSE), EA-6B Value Stream	Jacksonville	Florida	USA	Military
2007	Bronze	Robins Air Force Base, 572nd Commodities Maintenance Squadron, F-15 Wing Flight	Warner Robins	Georgia	USA	Military
2007	Shingo Prize	Hearth & Home Technologies	Mount Pleasant	Iowa	USA	Other
2007	Gold	Rock Island Arsenal, Joint Manufacturing and Technology Center, Forward Repair System (FRS) Value Stream	Rock Island	Illinois	USA	Military
2007	Silver	Rock Island Arsenal, Joint Manufacturing and Technology Center, Shop Equipment Contact Maintenance (SECM) System Value Stream	Rock Island	Illinois	USA	Military
2007	Shingo Prize	Raytheon Missile Systems	Louisville	Kentucky	USA	Military
2007	Gold	Fleet Readiness Center (FRC) East, H-1 Helicopter Line, MCAS	Cherry Point	North Carolina	USA	Military
2007	Shingo Prize	Baxter Healthcare, North Cove Plant	Marion	North Carolina	USA	Medical
2007	Bronze	Letterkenny Army Depot (LEAD), Power Generation Equipment	Chambersburg	Pennsylvania	USA	Military
2007	Finalist	BAE Systems Ground Division	York	Pennsylvania	USA	Defense Contractor

(Continued)

Year	Level	Company Name	City	State	Country	Industry
2007	Gold	Tobyhanna Army Depot, Antenna Transceiver Group	Tobyhanna	Pennsylvania	USA	Military
2007	Silver	Letterkenny Army Depot, Tactical Wheeled Vehicles, HMMWV	Chambersburg	Pennsylvania	USA	Military
2007	Shingo Prize	DENSO Manufacturing Tennessee, Inc., Instrument Cluster Division	Maryville	Tennessee	USA	Automotive
2007	Bronze	Corpus Christi Army Depot (CCAD), H-60 Pavehawk Joint Depot Level Maintenance (JDLM) Program	Corpus Christi	Texas	USA	Military
2007	Finalist	Boeing, KC 135 GATM	San Antonio	Texas	USA	Aviation & Aerospace
2007	Gold	Red River Army Depot, High Mobility Multipurpose Wheeled Vehicle (HMMWV) Center of Industrial and Technical Excellence	Texarkana	Texas	USA	Military
2007	Silver	Red River Army Depot (RRAD), Bradley Fighting Vehicle System (BFVS) Power Train Processes	Texarkana	Texas	USA	Military
2007	Silver	Red River Army Depot, Heavy Expanded Mobility Tactical Truck (HEMTT) Center of Industrial and Technical Excellence	Texarkana	Texas	USA	Military
2007	Silver	Hill Air Force Base, 574th Aircraft Structural Repair Squadron	Ogden	Utah	USA	Military
2006	Shingo Prize	Delphi Corporation, Delco Electronics, Plant 58	Chihuahua	Chihuahua	Mexico	Automotive
2006	Shingo Prize	Delphi Corporation, Plant 65	Queretaro	Queretaro	Mexico	Automotive
2006	Shingo Prize	Delphi Corporation, Plant 66	Queretaro	Queretaro	Mexico	Automotive

(Continued)

Year	Level	Company Name	City	State	Country	Industry
2006	Shingo Prize	TI Automotive, Mexico City Plant	Tultitlan	Tultitlan	Mexico	Automotive
2006	Shingo Prize	Methode Mexico, S.A. de C.V	Apodaca	Apodaca	Mexico	Other
2006	Shingo Prize	Steelcase, Inc.	City of Industry	California	USA	Distribution
2006	Shingo Prize	dj Orthopedics	Vista	California	USA	Healthcare
2006	Bronze	568th Fighter Avionics Squadron, 402d Maintenance Wing, Warner Robins Air Logistics Center, Robins Air Force Base	Warner Robins	Georgia	USA	Military
2006	Bronze	Robins Air Force Base, F-15 Programmed Depot Maintenance, Warner Robins Air Logistics Center	Warner Robins	Georgia	USA	Military
2006	Gold	Robins Air Force Base, C-5 Programmed Depot Maintenance	Warner Robins	Georgia	USA	Military
2006	Finalist	Rockwell Collins, Coralville Operations	Coralville	Iowa	USA	Aviation & Aerospace
2006	Gold	Rock Island Arsenal, Joint Manufacturing and Technology Center , Forward Repair System (FRS) Value Stream	Rock Island	Illinois	USA	Military
2006	Shingo Prize	Aspect Medical Systems	Newton	Massachusetts	USA	Medical
2006	Finalist	Freudenberg-NOK, Components Plant	Bristol	New Hampshire	USA	Automotive
2006	Bronze	Responsive Neutron Generator Product Deployment Center, Sandia National Laboratories	Albuquerque	New Mexico	USA	Military
2006	Bronze	Tobyhanna Army Depot, AN/TPS-75 Air Defense Radar System	Tobyhanna	Pennsylvania	USA	Military
2006	Silver	Letterkenny Army Depot, Tactical Vehicles – HMMWV Recap	Chambersburg	Pennsylvania	USA	Military

(Continued)

Year	Level	Company Name	City	State	Country	Industry
2006	Finalist	Scotsman Ice Systems, Fairfax Operation	Fairfax	South Carolina	USA	Other
2006	Finalist	John Deere, Power Products	Greeneville	Tennessee	USA	Consumer Goods
2006	Silver	Red River Army Depot, HMMWV Recap	Texarkana	Texas	USA	Military
2006	Gold	Hill Air Force Base, F-16 Aircraft Maintenance Squadron	Ogden	Utah	USA	Military
2006	Shingo Prize	Autoliv, Inc	Promontory	Utah	USA	Automotive
2006	Shingo Prize	Delphi Corporation, Milwaukee Operations	Oak Creek	Wisconsin	USA	Automotive
2005	Finalist	Delphi Automotive Systems Plant 32 RBE VII	Juarez	Juarez	Mexico	Automotive
2005	Finalist	Delphi Sistemas de Energia Plant 59	Torreon	Torreon	Mexico	Automotive
2005	Finalist	Noble Metal Processing de Mexico S. de R. L. de C. V. Silao Plant	Silao	Silao	Mexico	Automotive
2005	Shingo Prize	Delphi Corporation, Delphi Sistemas & Energy, Saltillo Operations Plant 39	Saltillo	Saltillo	Mexico	Automotive
2005	Shingo Prize	Delphi Ensamble de Cables y Componentes, Guadalupe II Plant 84	Guadalupe	Guadalupe	Mexico	Automotive
2005	Shingo Prize	Takata Seat Belts, Inc., Automortiz Americana, S.A.de C.V., Agua Prienta Plant	Agua Prienta	Agua Prienta	Mexico	Automotive
2005	Shingo Prize	Takata Seat Belts, Inc., Automortiz Americana, S.A.de C.V., Monterrey Plant 1	Apodaca	Apodaca	Mexico	Automotive
2005	Shingo Prize	Takata Seat Belts, Inc., Automortiz Americana, S.A.de C.V., Monterrey Plant 2	Apodaca	Apodaca	Mexico	Automotive

(Continued)

Year	Level	Company Name	City	State	Country	Industry
2005	Shingo Prize	Celestica de Monterrey	Monterrey	Monterrey	Mexico	Other
2005	Shingo Prize	The Boeing Company, Apache Longbow Program	Mesa	Arizona	USA	Aviation & Aerospace
2005	Gold	Robins AFB, C-5 Programmed Depot Maintenance	Warner Robins	Georgia	USA	Military
2005	Shingo Prize	BAE Systems, Platform Solutions	Fort Wayne	Indiana	USA	Aviation & Aerospace
2005	Finalist	Brazeway, Inc.	Adrian	Michigan	USA	Automotive
2005	Shingo Prize	Boston Scientific, Maple Grove Operations	Maple Grove	Minnesota	USA	Medical
2005	Shingo Prize	Hearth & Home Technologies	Lake City	Minnesota	USA	Other
2005	Shingo Prize	GDX Automotive	New Haven	Missouri	USA	Automotive
2005	Shingo Prize	The Boeing Company, Weapons ECC	St. Charles	Missouri	USA	Aviation & Aerospace
2005	Shingo Prize	Delphi Packard Electric, Vienna Molding Operations	Warren	Ohio	USA	Automotive
2005	Silver	Tinker AFB, Aircraft Division, KC-135 Branch	Tinker AFB	Oklahoma	USA	Military
2005	Shingo Prize	Lockheed Martin Missiles	Archbald	Pennsylvania	USA	Military
2005	Silver	Letterkenny Army Depot, Patriot Missile Air Defense System	Chambersburg	Pennsylvania	USA	Military
2005	Finalist	Takata Seatbelts, Inc. Del Rio Plant	Del Rio	Texas	USA	Automotive
2005	Shingo Prize	Autoliv, Inc	Tremonton	Utah	USA	Automotive
2005	Silver	Hill AFB, Commodities Branch, Pylons Shop	Ogden	Utah	USA	Military
2005	Silver	Hill AFB, F-16 Branch CCIP Business Unit	Ogden	Utah	USA	Military
2004	Finalist	Nemak Corp. of Canada Windsor Aluminum Plant	Windsor	Ontario	Canada	Automotive
2004	Shingo Prize	Delphi Corporation, CENTEC Plant 98	Ramos Arizpe	Ramos Arizpe	Mexico	Automotive

(Continued)

Year	Level	Company Name	City	State	Country	Industry
2004	Shingo Prize	Delphi Corporation, Delphi Sistemas & Energy, Chihuahua Operations	Chihuahua	Chihuahua	Mexico	Automotive
2004	Shingo Prize	Delphi Corporation, Electronics & Safety, Delnosa Operations, Plants 5 & 6	Reynosa	Reynosa	Mexico	Automotive
2004	Shingo Prize	Delphi Corporation, Plant 50	Del Parral	Del Parral	Mexico	Automotive
2004	Shingo Prize	Delphi Corporation, Plant 51	Casas Grandes	Casas Grandes	Mexico	Automotive
2004	Shingo Prize	Delphi Corporation, Plant 58	Meoqui	Meoqui	Mexico	Automotive
2004	Shingo Prize	Delphi Corporation, Tlaxcala Operations	Tlaxcala	Tlaxcala	Mexico	Automotive
2004	Finalist	GDX Automotive, Vehicle Sealing Products	Batesville	Arizona	USA	Automotive
2004	Finalist	Maytag-Searcy Laundry Products	Searcy	Arizona	USA	Consumer Goods
2004	Shingo Prize	Raytheon Missile Systems Mfg.	Tucson	Arizona	USA	Defense Contractor
2004	Shingo Prize	TI Automotive	Cartersville	Georgia	USA	Automotive
2004	Shingo Prize	ArvinMeritor	Columbus	Indiana	USA	Automotive
2004	Shingo Prize	Delphi Corporation, Electronics & Safety, Kokomo Operations, Plants 7 & 9	Kokomo	Indiana	USA	Automotive
2004	Finalist	Aspect Medical Systems, Inc.	Newton	Massachusetts	USA	Medical
2004	Finalist	Delphi Lansing Cockpit Plant	Lansing	Michigan	USA	Automotive
2004	Finalist	Delphi Saginaw Operations	Saginaw	Michigan	USA	Automotive
2004	Finalist	Delphi Saginaw Steering Plant 7	Saginaw	Michigan	USA	Automotive
2004	Finalist	GDX Automotive	New Haven	Missouri	USA	Automotive
2004	Finalist	TI Automotive Brake Tube Assemblies	Greeneville	Tennessee	USA	Automotive
2004	Shingo Prize	Maytag Jackson Dishwashing	Jackson	Tennessee	USA	Consumer Goods
2003	Shingo Prize	Delphi Corporation, Delco Electronics, Delnosa Operations Plant 1-4	Reynosa	Reynosa	Mexico	Automotive

(Continued)

Year	Level	Company Name	City	State	Country	Industry
2003	Shingo Prize	Delphi Corporation, Juarez Operations	Juarez	Juarez	Mexico	Automotive
2003	Shingo Prize	Lockheed Martin Aeronautics, F-117 Plant	Palmdale	California	USA	Defense Contractor
2003	Shingo Prize	Medtronic Xomed	Jacksonville	Florida	USA	Medical
2003	Shingo Prize	The HON Company	Cedartown	Georgia	USA	Consumer Goods
2003	Shingo Prize	Affordable Interior Systems	Hudson	Massachusetts	USA	Consumer Goods
2003	Shingo Prize	Delphi Corporation, Delco Electronics	Flint	Michigan	USA	Automotive
2003	Shingo Prize	TI Automotive	Caro	Michigan	USA	Automotive
2003	Shingo Prize	TI Automotive	New Haven	Michigan	USA	Automotive
2003	Shingo Prize	Vibracoustic	Manchester	New Hampshire	USA	Automotive
2003	Shingo Prize	Symbol Technologies; Holtsville, NY, McAllen, TX & Reynosa, Mexico	Holtsville	New York	USA	Computer & Electronic
2003	Shingo Prize	Delphi Corporation, Packard Electric, Warren Plant 19	Warren	Ohio	USA	Automotive
2003	Shingo Prize	Autoliv, Airbag Module Facility	Ogden	Utah	USA	Automotive
2003	Shingo Prize	Autoliv, Inflator Facilities; Brigham City and Ogden, UT	Brigham City	Utah	USA	Automotive
2003	Shingo Prize	Merillat Industries	Atkins	Virginia	USA	Other
2002	Shingo Prize	Delphi Corporation, Alambrados y Circuitos Electricos VII	Los Mochis	Los Mochis	Mexico	Automotive
2002	Shingo Prize	Delphi Corporation, Delco Electronics de Mexico, Deltronicos	Matamoros	Matamoros	Mexico	Automotive
2002	Shingo Prize	Delphi Corporation, Harrison Thermal Systems, Rio Bravo XX	Juarez	Juarez	Mexico	Automotive
2002	Shingo Prize	Ford Motor, Chihuahua Engine	Chihuahua	Chihuahua	Mexico	Automotive
2002	Shingo Prize	Grupo CYDSA, IQUISA; Monterrey and Tlaxcala, MX	Monterrey	Monterrey	Mexico	Chemical

(Continued)

Year	Level	Company Name	City	State	Country	Industry
2002	Shingo Prize	Tyco Fire and Security, Sensormatic	Puerto Rico	Puerto Rico	Puerto Rico	Other
2002	Shingo Prize	Lockheed Martin, Missile and Fire Control, Pike County	Troy	Alabama	USA	Defense Contractor
2002	Shingo Prize	Woodland Furniture	Idaho Falls	Idaho	USA	Consumer Goods
2002	Shingo Prize	Ford Motor, Chicago Assembly	Chicago	Illinois	USA	Automotive
2002	Shingo Prize	Vibration Control Technologies	Ligonier	Indiana	USA	Automotive
2002	Shingo Prize	Freudenberg-NOK	Shelbyville	Indiana	USA	Aviation & Aerospace
2002	Shingo Prize	Ensign-Bickford; Graham, KY, Simsbury, CT, Spanish Fork, UT, & Wolf Lake, IL	Graham	Kentucky	USA	Other
2002	Shingo Prize	Bridgewater Interiors	Detroit	Michigan	USA	Automotive
2002	Shingo Prize	Delphi Corporation, Adrian Operations,	Adrian	Michigan	USA	Automotive
2002	Shingo Prize	Ford Motor, Romeo Engine	Romeo	Michigan	USA	Automotive
2002	Shingo Prize	Delphi Corporation, Cortland Molding	Cortland	Ohio	USA	Automotive
2002	Shingo Prize	Bridgestone/Firestone	Aiken County	South Carolina	USA	Automotive
2001	Shingo Prize	Ford Motor, Essex Engine	Windsor	Ontario	Canada	Automotive
2001	Shingo Prize	Baxter Healthcare	Mountain Home	Arkansas	USA	Healthcare
2001	Shingo Prize	Freudenberg-NOK	Cleveland	Georgia	USA	Aviation & Aerospace
2001	Shingo Prize	Benteler Automotive, Hagen Exhaust	Grand Rapids	Michigan	USA	Automotive
2001	Shingo Prize	Ford Motor, Ohio Assembly	Avon Lake	Ohio	USA	Automotive
2001	Shingo Prize	Johnson Controls	Greenfield	Ohio	USA	Computer & Electronic
2000	Shingo Prize	Ford Motor, Windsor Engine	Windsor	Ontario	Canada	Automotive
2000	Shingo Prize	Delphi Corporation, Componentes Mecanicos	Matamoros	Matamoros	Mexico	Automotive
2000	Shingo Prize	Grupo CYDSA, Policyd La Presa	Tlanepantla	Tlanepantla	Mexico	Automotive

(Continued)

Year	Level	Company Name	City	State	Country	Industry
2000	Shingo Prize	Freudenberg-NOK	LaGrange	Georgia	USA	Aviation & Aerospace
2000	Shingo Prize	Delphi Corporation, Steering Plant 6	Saginaw	Michigan	USA	Automotive
2000	Shingo Prize	Baxter Healthcare	Marion	North Carolina	USA	Healthcare
2000	Shingo Prize	Lucent Technologies, Product Realization Center	Mt. Olive	New Jersey	USA	Communication
2000	Shingo Prize	Lockheed Martin Aeronautics	Fort Worth	Texas	USA	Aviation & Aerospace
1999	Shingo Prize	Delphi Corporation, Rimir	Matamoros	Matamoros	Mexico	Automotive
1999	Shingo Prize	Grupo CYDSA, Sales del Istmo	Coatzacoalcos	Coatzacoalcos	Mexico	Automotive
1999	Shingo Prize	Spicer Cardanes	Queretaro	Queretaro	Mexico	Automotive
1999	Shingo Prize	Wiremold	West Harford	Connecticut	USA	Computer & Electronic
1999	Shingo Prize	O.C. Tanner	Salt Lake City	Utah	USA	Other
1999	Shingo Prize	Federal-Mogul	Hampton	Virginia	USA	Automotive
1998	Shingo Prize	Tremec Transmisiones Y Equipos Mecanicos	Queretaro	Queretaro	Mexico	Automotive
1998	Shingo Prize	Grupo CYDSA, IQUISA	Coatzacoalcos	Coatzacoalcos	Mexico	Chemical
1998	Shingo Prize	Freudenberg-NOK, Gasket Lead Center	Manchester	New Hampshire	USA	Aviation & Aerospace
1998	Shingo Prize	Coach	Carlstadt	New Jersey	USA	Textiles & Apparel
1998	Shingo Prize	Johnson Controls	Lexington	Tennessee	USA	Computer & Electronic
1998	Shingo Prize	Lear	Winchester	Virginia	USA	Automotive
1998	Shingo Prize	Milwaukee Electric Tool	Brookfield	Wisconsin	USA	Other
1997	Shingo Prize	Grupo CYDSA, Policyd	Altamira	Altamira	Mexico	Automotive
1997	Shingo Prize	Industrias CYDSA Bayer	Coatzacoalcos	Coatzacoalcos	Mexico	Automotive
1997	Shingo Prize	Johnson Controls, ASG, Quality Drive	Georgetown	Kentucky	USA	Automotive
1997	Shingo Prize	TechnoTrim	Livonia	Michigan	USA	Automotive
1997	Shingo Prize	Champion International	Sartel	Minnesota	USA	Consumer Goods

(Continued)

Year	Level	Company Name	City	State	Country	Industry
1997	Shingo Prize	Johnson Controls, ASG	Jefferson City	Missouri	USA	Automotive
1997	Shingo Prize	Perfecseal—Bemis Company	Philadelphia	Pennsylvania	USA	Healthcare
1997	Shingo Prize	Johnson Controls, ASG	Linden	Tennessee	USA	Automotive
1997	Shingo Prize	Johnson Controls, ASG	Pulaski	Tennessee	USA	Automotive
1996	Shingo Prize	Eaton Yale Ltd.	St. Thomas	Ontario	Canada	Automotive
1996	Shingo Prize	Johnson Controls, ASG	Orangeville	Ontario	Canada	Automotive
1996	Shingo Prize	Johnson Controls, ASG, FoaMech	Georgetown	Kentucky	USA	Automotive
1996	Shingo Prize	Ford Motor, Cleveland Engine Plant #2	Cleveland	Ohio	USA	Automotive
1996	Shingo Prize	Merix	Forest Grove	Oregon	USA	Distribution
1996	Shingo Prize	Ford Motor, North Penn Electronics	Lansdale	Pennsylvania	USA	Automotive
1996	Shingo Prize	Harris Corporation, Farinon Division	San Antonio	Texas	USA	Communication
1995	Shingo Prize	Nucor—Yamato Steel	Blytheville	Arkansas	USA	Building Material
1995	Shingo Prize	LifeScan—Johnson & Johnson	Milpitas	California	USA	Medical
1995	Shingo Prize	The Foxboro Company, I/A Division	Foxboro	Massachusetts	USA	Non-Profit
1995	Shingo Prize	MascoTech—Braun	Detroit	Michigan	USA	Automotive
1995	Shingo Prize	Vintec	Murfreesboro	Tennessee	USA	Automotive
1995	Shingo Prize	Tennalum, Kaiser Aluminum	Jackson	Tennessee	USA	Aviation & Aerospace
1994	Shingo Prize	Ford Electronics	Markham	Ontario	Canada	Automotive
1994	Shingo Prize	Union Carbide, Ethyleneamines Business; Danbury, CT, Taft, LA, & Texas City, TX	Danbury	Connecticut	USA	Chemical
1994	Shingo Prize	Johnson & Johnson, Medical Vascular Access	Southington	Connecticut	USA	Medical
1994	Shingo Prize	AT&T Technologies, Microelectronics	Orlando	Florida	USA	Communication

(Continued)

Year	Level	Company Name	City	State	Country	Industry
1994	Shingo Prize	Lucent Technologies, Microelectronics Group	Orlando	Florida	USA	Communication
1994	Shingo Prize	General Tire	Mt. Vernon	Illinois	USA	Automotive
1994	Shingo Prize	The Timken Company	Gaffney	South Carolina	USA	Automotive
1994	Shingo Prize	Alcatel Networks Systems; Richardson, TX, Longview, TX, Raleigh, NC, Clinton, NC, & Nogales, Mexico	Richardson	Texas	USA	Communication
1993	Shingo Prize	Gates Rubber	Siloam Springs	Arkansas	USA	Automotive
1993	Shingo Prize	Wilson Sporting Goods, Golf Balls	Humboldt	Tennessee	USA	Consumer Goods
1992	Shingo Prize	AT&T Technologies, Microelectronics, Power Systems	Mesquite	Texas	USA	Communication
1992	Shingo Prize	Lucent Technologies, Microelectronics Group, Power Systems	Mesquite	Texas	USA	Communication
1992	Shingo Prize	Iomega	Roy	Utah	USA	Computer & Electronic
1991	Shingo Prize	Glacier Vandervell	Atlantic	Iowa	USA	Automotive
1991	Shingo Prize	Lifeline Systems	Watertown	Massachusetts	USA	Medical
1991	Shingo Prize	Dana Mobile Fluid Products Division	Minneapolis	Minnesota	USA	Automotive
1991	Shingo Prize	Exxon Chemical, Butyl Polymers Americas	Houston	Texas	USA	Chemical
1990	Shingo Prize	United Electric Controls	Watertown	Massachusetts	USA	Distribution
1989	Shingo Prize	Globe Metallurgical	Beverly	Ohio	USA	Chemical

RECENT SHINGO PRIZE RECIPIENT PROFILE SHEETS

Introduction

Corporate Overview
Ball Corporation is a leading global beverage can maker with a strong vision to make the can the most sustainable package in the beverage supply chain. To support this, Ball has a program in place called "Drive for 10," which is a mindset around perfection with a great sense of urgency around future success.

These are the three pillars of Drive for 10:
- We Know Who We Are
- We Know Where We Are Going
- We Know What is Important

As a global entity, Ball operates in 120 locations around the world, of those more than 70 are beverage packaging plants including joint ventures. The beverage packaging business has a volume of around 110 billion units a year and a presence in 30 countries with an excess of 12,000 employees in the beverage packing regions alone.

Ball's job is to make those cans as efficiently, profitably and sustainably as possible so that customers can attract and retain the loyalty of the people who consume their products. This is why, wherever you go in Ball, you will see a common focus on operational excellence, innovation and safety to meet stakeholder's expectations.

The vision at Ball is "We want to be close to customers and win more in the market."

The Naro Fominsk Facility
The ends making facility (NARE), part of the Beverage Packaging Europe (BBPE) region of Ball, is in the town of Naro Fominsk, one hour southwest of Moscow. Construction started in July 2003 with Modules 1 and 2 delivering customer ends in early 2004.

NARE is one of four Ball end manufacturing plants in Europe, with 153 employees capable of producing six billion ends annually. The 5846-sq.m. facility is divided into four "Modules" (continuous flow processing lines) including six high-speed stamping presses and a high level of automation on three of the four lines. Key customers include Coca-Cola, Pepsi and Carlsberg. NARE operates a 24/7 pattern of work where customer's expectations drive them to perform at their very best every day.

Product and Process
The facility produces just one end size (202) but with two designs, Standard 202 and CDL 202. Through a combination of customer requirements including colored metal and laser engraving, there are more than 28 SKUs. The process has four recognizably phases: cutting and shaping from rolled aluminium stock into basic disc ends, sealing compound application, shaping and cutting with tab (ring pull) attachment, and finished ends packed onto pallets ready for dispatch.

A thorough approach is taken toward safety and training to deliver a healthy work environment and great customer service. Production is carried out and controlled using the concept of "quality at the source" which gives a focus on the quality of the process rather than the product's characteristics.

Continuous Improvement
In 2000 NARE became part of a global Lean journey based on a specific set of tools and systems: culture, VSM, 5S, TPM, SMED and Six Sigma. From 2004, progress has been annually reviewed in the form of an internal assessment and performance was awarded as a prize of Bronze, Silver or Gold. The factory achieved Gold within four years, three in a row from 2007 to 2010 and again in 2012.

In 2011, the *Shingo Model* was introduced with an internal assessment where the facility was the first end plant in Europe to be chosen for the challenge.

Ball involves shop floor employees on a daily basis in identifying improvements and encourages them to think systemically and scientifically. By sharing the strategic view, helping employees with resources, providing more freedom in making decisions and getting their feedback on ways to improve the factory; Ball believes they will be successful in achieving the vision of being the best end making facility in BBPE.

Every idea is very important; each kaizen event moves closer to perfection. Every year Ball's strategic plan includes safety kaizen events, and for the last two years they have focused on people and behavior to reduce near misses. This approach is a major contributor in continuing the factory with over two thousand lost time accident free days.

As a part of the improvement process Ball has developed and embedded a quality improvement program named Behavior Based Quality (BBQ). This approach was born in this facility and is shared with the other end plants across BBPE.

BALL BEVERAGE PACKAGING EUROPE, Naro Fominsk Ends   Naro Fominsk, Russia

Noteworthy Achievements for our Factory
- 2,290 days without a lost time accident (LTA) (Feb '16)
- When benchmarked against BBPE ends facilities, ranked top in eight categories: OEE, efficiency, line spoilage, HFI spoilage, total spoilage, MTBF, NOC and maintenance spend/000 ends
- Company People Award in 2010, again in 2015

Safety & Environment
- Zero LTAs for six consecutive years since 2010
- Company Safety Award for passing 1,000 and 1,500 days since 2009 for no LTAs; 2,000 days in late 2015. Top three performance in BBPE
- Electric power consumption reduced by 25.8% over the last 5 years
- Evacuation drill training provided to all employees annually
- Waste reduction: Aluminum coil packaging waste reduction activities were agreed by both the plant and metal supplier. Now receiving lighter packaging with no plastic cover.
- Behavior Based Safety (BBS) Program in its third year and benchmarked across the company
- Blind area traffic light warning system on walkways
- Fork trucks equipped with blue spot warning light

Quality
- British Retail Consortium (BRC) accreditation since 2011
- Facility has trained 'technicians' with technical skills in Six Sigma, 5S, TPM and SMED
- Return pallet per billion ends produced reduced to 0 by 2015; benchmark for end plants across BCE
- NOC per billion produced reduced by 89% between 2010 and 2015; benchmark for end plants in BCE
- Digital inspection camera systems installed on all conversion presses
- Real-time visual system and downtime management in operation
- Real-time quality system management and statistical control
- 3D Microscope: Score assessment completed quickly and precisely
- Behavior Based Quality (BBQ) program in place
- Company Quality Audit 94.4% in 2015; benchmark for end and can plants across BCE

Employee Morale
- Absenteeism and employee turnover has reduced by 0.52% and 0.58% from 2010 and 2011 respectively
- Health and well-being lifestyle promotions include healthy menu at canteen
- Home safety week events, and a safety focus at internal conferences
- Safety leaflets for employees' children «safe behavior during winter holidays» in New Year sweet package gifts
- 100% vacancies are closed by internal pool of candidates in accordance with succession plan
- Two annual 'Blue Chip' recognition awards for employees
- Employee attestation system in place – in 2015, 100% employees reached max level
- 'Best Crew' evaluation conducted on a monthly basis, awarded twice a year

Cost Reductions
- Efficiency has increased by 8.3%
- Conversion cost/1000 ends made has reduced by 30% since 2010
- 94% reduction in spoilage from 2010 to 2015; benchmark for end plants in BBPE

People

Ball's mission, always, is to develop people into leaders that can take them to the next level in their journey to becoming the best end maker in the world. Employees are not just seen as a source of labor, but as process experts in the continuous improvement process.

The Naro Fominsk location is often used as a training base for all company Russian plants. Currently black belt training is in progress on-site for four Russian plants, with an added bonus for NARE to learn new things from those visiting the plant.

For more information contact:
Ray Howcroft
Lean Enterprise Manager
Direct Line +44 (0) 1924 834017
Mobile +44 (0)7766 160993
Email: Ray.Howcroft@ball.com

ThermoFisher
SCIENTIFIC

Corporate Overview
Thermo Fisher Scientific Inc. is the world leader in serving science, with revenues of $18 billion and approximately 55,000 employees globally. Thermo Fisher Scientific's mission is to enable customers to make the world healthier, cleaner and safer. They help customers accelerate life sciences research, solve complex analytical challenges, improve patient diagnostics and increase laboratory productivity. Through premier brands – Thermo Scientific, Applied Biosystems, Invitrogen, Fisher Scientific and Unity Lab Services – Thermo Fisher Scientific offers an unmatched combination of innovative technologies, purchasing convenience and comprehensive support.

Vilnius Site Overview
Thermo Fisher Scientific's Vilnius site has world-class capabilities in manufacturing products for the life science research market, specifically in molecular, protein and cell biology, with an outstanding research and development (R&D) center, focused on the development of new products in all aspects of molecular, protein and cell biology. The facility employs over 700 people in a variety of roles including 100 researchers, making the Vilnius site one of the largest private R&D centers in the whole region.

Thermo Fisher Scientific's products are broadly used worldwide to study gene structure, expression and genetic variation, and to create new diagnostics methods for congenital, hereditary and infectious diseases. All products are manufactured with constant focus on quality and lead times in six value streams using the flow approach. Clean room facilities from ISO 5 to ISO 8 (in total 20 k sq. ft.) help to ensure high quality and lot-to-lot consistency of products. The business is constantly growing: the number of employees increased from 420 in 2013 to 700 in 2017, and 220 k sq. ft. in three buildings.

Mission, Strategy and Behavior
The mission is supported by the site strategy statement: "All products in 24 hours with zero waste." It is visible throughout the site for all employees and everybody is working to achieve it. The everyday motto for all employees is: Job = Daily Work + Kaizens.

An educated employee community (>80% with university degree, 8% with PhD) shows commitment to learning and continuous improvement and understands how critical the link is between the company strategic goals and employee personal objectives. With that in mind, the site developed a unique process for strategy deployment (based on X-Matrix and strategic A3 tools). According to employee survey results, 98% of all employees understand the site's strategic goals and know how they contribute to achieve these goals by their daily work.

Company Profile Sheet

Continuous Improvement Process
The foundation for the site achievements for the last three years is the scientific PDSA (Plan-Do-Study-Act) approach used on a daily basis via kaizen, just do it, or PPI (Practical Process Improvement – structured continuous process improvement business system) classic projects. The site has implemented more than 1500 kaizens during the last three years. The PPI Business System includes both the 8-step methodology based on PDSA scientific method and all Lean management and manufacturing tools.

Quality Management Systems
- ISO 9001:2008 Quality Management System in 1996, last recertification in 2016.
- ISO 13485:2003 Quality Management System for medical devices in 2010, last recertification in 2016.
- ISO 14001:2004 Environmental management system in 2003, last recertification in 2015.
- OHSAS 18001 Employees Health and Safety Management System in 2015.
- US Code of Federal Regulations (21CFR part 820) registered in 2016.

Customer and Quality
- World Class Customer Allegiance Score (CAS) is > 70 starting from 2013.
- World Class Line Items Fill Rate (LIFR) is > 99%, on-time delivery is > 98%, a result of joint kaizens with suppliers, carriers and distribution centers.
- Number of customer complaints was reduced by 33% from 2013 to 2016. Each customer complaint is analyzed using RCA (Root Cause Analysis). >90% of all processes are standardized, therefore each abnormal condition or non-conformity is obvious.
- Orders for catalog products are fulfilled in 24 hours from order received. The site is one of the most advanced organizations in the Life Science industry when measured by the lead times with constant focus on manufacturing lead times reduction through flow and innovative technologies implementation.
- Sales, marketing, R&D, quality employees visit customers on a regular basis. Days of Science are held annually to strengthen relationships with the customers.
- Internal Customer Week: an event is held each year in order to define the internal customers within the site and to set commitments between departments and groups for the next year.

Safety & Environmental
- Over 1000 safety days currently. Reporting about health, environmental and security near misses, hazardous identifications is performed on a daily basis during stand-up meetings. Monthly awards for the best hazardous identifications are given for employees.

ThermoFisher
S C I E N T I F I C

Company Profile Sheet

- Annual safety and first aid training are organized for employees.
- Safety gemba walks performed by all managers on a daily basis.
- The Greenest Enterprise of the Year in 2014 by the leading national business newspaper.

Employees

- Employees Involvement Index with 81% in 2016, engaged, informed and empowered employees contribute to customers' product experience and satisfaction.
- The Highest Reputation Index in Lithuania according to the National Business Reputation Survey in 2017 and 2016. According to the National Business Reputation Survey 2017, Thermo Fisher has the highest reputation index in Lithuania for the second year in a row. Moreover, the Reputation Index has increased from 74 to 81.
- According to employee survey results 98% of all employees understand the site's strategic goals.
- 90% of all employees are trained on the 8-step method (PDSA) and Lean tools; training on the error prevention system was also delivered to 90% of all employees in 2016.
- >90% of all employees are involved in PPI activities each quarter.
- 90% of managers do gemba walks, the number of ideas increased from 240 in 2013 to 5002 in 2015.
- Woman Employee Resource Group (WERG) is focused on career development and social initiatives and is open for all employees independent of their demographic characteristics.
- Each year over 50% of employees volunteer in Corporate Social Responsibility (CSR) activities.

Outstanding Reputation in the Region

- The highest reputation index in Lithuania according to the Reputation Study 2017 and 2016.
- Investor of the Year by the leading business daily in 2016.
- Algimantas Markauskas (General Manager) honored as CEO of the Year in 2015 and 2013.
- The Most Responsible Enterprise of the Year in 2015 by the Investors' Forum.
- Employer of the Year in Lithuania in 2014 and 2012 by the Ministry of Social Security and Labor.
- Award of Socially Responsible Business in 2013 by Ministry of Social Security and Labor.
- The Most Innovative US company in Lithuania in 2013 by the American Chamber of Commerce.
- Social Innovator of the Year in 2013 by the Ministry of Economy.

Collaboration with Universities

- Internship program: >40 scholarships for the students to do final BSc/MSc thesis research in R&D and manufacturing.
- Participation in development of study programs.
- Lectures at universities: 8 employees as part-timers.
- Delivering lectures on Lean for MSc students.
- Mobile Bioclass, the first country traveling biosciences lab:
  - Promoting bioscience studies among high school students.
  - >200 high schools visited in 90 cities and towns.

For more information, contact:

Alina Stura
Senior Business Excellence Specialist
Thermo Fisher Scientific's Vilnius site
Phone: +370 5250 7913
Email: alina.stura@thermofisher.com

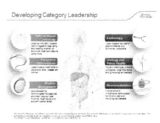

Developing Category Leadership

Boston Scientific Corporate Overview

Boston Scientific (BSC) transforms lives through innovative medical solutions that improve the health of patients around the world. As a global medical technology leader for more than 35 years, BSC advances science for life by providing a broad range of high-performance solutions that address unmet patient needs and reduces the cost of healthcare.

Focusing on patient care enables BSC to achieve their vision of being the highest performing global medical solution provider. They do this by having patient care at the center of all activities.

BSC Guiding Priorities

The company's 2015 revenue was $8 billion with $876 million invested in research and development. Boston Scientific has 24,000 employees around the world, a sales forces in more than 40 countries, and 11 manufacturing facilities worldwide. The company has more than 13,000 products in its portfolio, over 16,000 patents issued and approximately 7,200 applications pending worldwide. The Boston Scientific six core values – meaningful innovation, caring, high performance, global collaboration, diversity and winning spirit – clearly express the company's corporate philosophy.

Boston Scientific Cork

In 1997, Boston Scientific acquired a 30-acre site in Cork, Ireland, that included a 25,000 ft^2 facility featuring a historic protected building, the Munster Institute (20,000 ft^2). The Cork manufacturing facility supports the Boston Scientific Peripheral Interventions, Endoscopy, Interventional Cardiology and Urology and Pelvic Health businesses. The site has an annual output of 5.5 million units per year, works a two-shift operation over a five-day week and has a workforce of 800 employees of whom 570 are product builders and the rest work in support roles.

Products

Boston Scientific Cork manufactures a diversified portfolio of medical devices for distribution across the globe. The product portfolio includes active and access catheters, occlusion coils and microspheres, inflation devices and atherectomy devices. As a worldwide manufacturer and distributor of medical devices, the Cork facility operates under a Global Quality System. This System addresses the management responsibility, resource management, product realization, measurement, analysis and improvement of products in fulfilment of all applicable regulations and standards.

The Cork site's quality policy is "I Improve the Quality of Patient Care and all things Boston Scientific" and every employee takes ownership for quality and understands how his or her performance directly impacts patient care.

Site Vision and Strategy

The Cork site's vision is "To be the Highest Performing Global Medical Solution Provider" by focusing on patient care, as defined by guiding priorities. The strategic quality process (SQP) is a key driver in how they run the business. The SQP enables them to prioritize, execute and monitor activities and align resources accordingly. The site balanced scorecard aligns the team to the SQP and forms the basis of core team metrics. The reward and recognition of employees underpins the process.

Ideal Behaviors

Ideal behaviors were developed by teams within Cork and are reinforced by this philosophy: "I want to be here... I do my best work here."

The site's ideal behaviors are closely aligned to the Boston Scientific Global Core Values: Caring, Diversity, Global Collaboration, High Performance, Meaningful Innovation and Winning Spirit. Together, they reinforce an open and honest work environment where trust and respect prevails thereby allowing engagement at all levels of the organization to the benefit of all.

Continuous Improvement

Continuous improvement has been at the core of the Cork site's culture since its inception. Over time, the team has implemented a number of programs and initiatives to strengthen our focus on continuous improvement at all levels. For instance, the team adopted Lean as a formal strategy in 2001 to improve production and business processes. The Lean tools and systems have become the standard for doing business and have evolved and improved over time as the business has grown. In 2012, Boston Scientific introduced Shingo as a platform to drive cultural change and to engage and involve everyone in the future of the business. With a daily focus on team huddles, tier meetings, leaders standard work, continuous improvement and problem solving, core and functional teams have had the tools and opportunities to identify and rectify critical issues in their areas more quickly and efficiently.

The Cork team is continually training employees deeper in the organization on skills such as problem solving so they too can have a more active role in finding improvement opportunities. Additionally, the continuous improvement culture benefits by the Value Improvement Process (VIP) that drives process improvements where cost reduction, product quality enhancement, customer responsiveness and waste elimination are the goals. Together, these programs and activities allow them to identify, prioritize and execute projects that create value for customers. They regularly celebrate the continuous improvement culture at the annual Recognize Success Fair, which has been in place for 17 years.

Customers

Boston Scientific is committed to offering its customers the deepest, most innovative and cost-effective portfolio of products and solutions in the market. Providing solutions to today's healthcare challenges requires them to develop a portfolio that is shaped by the convergence of differentiated technological expertise, external collaboration opportunities and market needs. As the company looks to the future, the products and programs they have brought to market through organic development and acquisitions, complemented by an innovative pipeline and continued global expansion will benefit even more patients. They are focused on meeting the needs of patients and customers by delivering the results expected of a high-performance company. Boston Scientific offers the opportunity for all employees to view live cases regularly to provide insight into the impact and functionality of the products in patients' lives.

People and Environment

Boston Scientific promotes professional and personal safety in all aspects of the business in order to create an environment that is safe. Safety is always the first topic to be discussed at daily team meetings, emphasizing its importance as a first priority supporting the philosophy "safety first, safety always." They have a behavioral safety program that serves as the internal compass and method of preventing accidents.

The Cork facility is a "zero landfill" site. They segregate waste for recycling, and send the balance of the waste to a facility for energy generation. They have achieved 10 consecutive years of fully compliant ISO14001 environmental audits. The success of the environmental management system is based on strong integration of the ISO14001 environmental management system into all aspects of the operation. They established a Green Team in 2013 to promote sustainable practices at work, at home, and in the community by fostering and encouraging a deeper understanding and concern for natural surroundings.

The Corporate Social Responsibility (CSR) efforts include supporting local charities. Employees nominate a number of charities per year and one is selected to be the Charity Partner each year based on an open-voting system.

The CSR also includes programs such as STEM (Science, Technology, Engineering and Mathematics), Junior Achievement, a scholarship program, employee well-being, and sports and social clubs.

Boston Scientific Cork Achievements:

- Shingo Prize recipient 2016
- Grad Ireland Award in 2015 (BSC Ireland)
- 5.5m devices manufactured per year
- 6 patients' lives impacted per minute
- >7,000 years of tacit knowledge
- 60% roles filled with internal applicants over the past 12 month
- >30 employees embark on further education annually
- Over €50k invested on local scholarship programs
- €500k contributed to local charities over the past 10 years

For more information, please contact:
Theresa Moloney
Boston Scientific
Cork, Ireland
Tel: +353 21 4531638

Introduction

Corporate Overview

Rexam is a leading global beverage can maker with sales in the region of $6.5 billion in 2014. With 55 can making plants in more than 20 countries, Rexam employs around 8,000 people and is headquartered in London, England. As an established member of the FTSE 100, Rexam is a business partner to some of the world's most famous and successful consumer brands such as Coca-Cola, Red Bull, Heineken, and PepsiCo.

Rexam's vision to "be the best global consumer packaging company" is strongly supported by five embedded core values: Continuous improvement that is always seeking perfection; Trust meaning loyalty to each other in a respectful way and commitment to customers, community and the environment; Teamwork to get the best outcome, together; Recognition to the most important asset, people; and Safety - recently named in Rexam's five values and consistent with respecting every individual.

They have been committed to a path of Lean enterprise across all operations as a way of putting values into practice for over ten years. This approach is an essential part of what is called the "Rexam Way."

The Rexam México Facility

Rexam Beverage Can Americas S. A. de C.V is based in Querétaro, Mexico. The operation was formed as a joint venture in September 1995 by the partners Vitro and Rexam till October 2004 when Rexam acquired 100% of the assets, supplying Mexico, Central America and the Caribbean area.

Of the 159 people on site, there are 140 involved in plant operations and management, and 19 people are responsible for finance, logistics and sales. The 520,000-sq. ft. facility has three aluminum can manufacturing lines, two for 12 oz. and one for 24 oz. sizes dedicated to customers such as Coca-Cola, Heineken, Hornell, and Modelo, with a total production of 3.9 million 12 oz. cans and 1.1 million 24 oz. cans each day. The plant operates 24 hours a day, seven days a week, and relentlessly works to exceed customer expectations.

Product and Process

The site is capable of producing two size beverage cans – 12 oz. and 24 oz. – for 18 customers, resulting in over 200 active label options (designs). The ability to continuously improve changeover times to meet requirements for smaller lot sizes and make faster deliveries is at the heart of maintaining a competitive advantage.

To make a beverage can, the key process steps involve stamping and drawing aluminum discs from rolled stock to form a can shape. The cans are then washed before being labelled using in-line screen printers. They are then shipped to customer filling locations across Mexico and Central America.

Speed and quality are the key drivers of process performance and OMS and QAS (Statistical Quality System); manufacturing operations management and enterprise manufacturing intelligence provide real-time production data and process monitoring capability.

Continuous Improvement Process

Rexam has operated a global system of Lean enterprise since 2003, structured around building capability in specific tool sets of culture, VSM, 5S, TPM, SMED and Six Sigma.

Rexam's best plants were invited to participate in a more sophisticated assessment known as "Beyond Gold," where the implementation of lean systems is appraised. A system of merit is used to recognize achievement starting at Emerald, then Sapphire, and finally Diamond. In recognition of Lean efforts, this facility became the first plant in Rexam North America to be awarded "Beyond Gold" diamond level twice.

Rexam México Achievements

- 2,228,989 work hours without lost time accident at the end of May 2015
- Lean enterprise achievements:
  - Silver 2004
  - Gold 2005, 2006, 2008, 2009, 2010
  - Beyond Gold Sapphire in 2011
  - Beyond Gold Diamond in 2012
- Rexam can designs are recognized worldwide for their print quality. (Worldstar Awards 98, 00, 01, 02, 03, 04, 05 and 06), and for innovation in 2010. Since 2011 to 2015 this plant was recognized in the industry's Can Maker magazine and received second place in innovation in 2012.
- Clean Industry Certification since 2001

Safety & Environmental

- Zero lost time accidents (LTA) since Sept. 2010.
- Rexam Chief Executive Officer Safety Award for

achieving one million hours with zero lost time incidents in 2009 and 2013
- Clean Industry Certification since 2011, demonstrating sustained compliance with Mexican environmental legislation
- Benchmarked against North and Central America Rexam plants, ranked top three in safety audit
- Annual safety training for all employees
- Monthly safety visits from management covering all areas of operations, warehousing and administrative areas
- 100% participation from operative and staff personnel in behavioral-based safety program.
- Zero environmental accidents
- Maintaining a program that helps to promote the recycling of aluminum cans open to the public
- Electricity consumption reduced by 25% between 2011 and 2014
- Gas consumption reduced by 14% between 2014 and 2014

Quality
- 2012 FSCC-22000 Food Safety accreditation
- Two black belts (plant manager, assistant plant manager)
- Six senior team members are green belts
- QAS (quality assurance systems) and QAS Minitab real-time SPC process monitoring covers 100% of production
- Number of customer complaints has reduced from 81 in 2013, to 76 in 2015, for a reduction of 38% over those years
- Coca-Cola, Heineken, Modelo Group and Omnilife trust supplier
- Real-time visual plant manufacturing system and downtime management (OMS)
- Automated in-line vision systems to detect mix label and flange-necker defects
- Support customers on location in order to improve their can line "run-ability"
- Between 2011 and 2015, five green belt projects including support areas, with participation from the people of the plant

Employee Morale
- Rotation below 1.60% since 2010, actual 2015 1.25%
- Absenteeism rate from 0.05% in 2013 to 0.03% in 2014, up-to-date 2015 0.02%
- Engagement score from 69% in 2012 to 74% in 2013

- 81% are extremely satisfied to work with REXAM as best place to work
- 21 employees with more than 20 years of service, 17 with more than 15 and 20 with more than 10 years
- Outdoor teambuilding sessions to learn and adopt principles and behaviors related with Rexam and Shingo culture
- High school program for those who desire to conclude their studies
- English programs for people who need it for their development and job requirements
- 14 operators recognized as qualified and trained for development of their skills in 2015
- Qualified and expert training process for operations positions
- Rexam Blue Chip program in place to recognize outstanding individual effort

Delivery Performance and Cost Reductions
- Percentage of deliveries made on-time and in-full have been maintained in levels close to 100% since 2000
- OEE 12 oz. has increased from 80.10% to 84.41% and OEE 24 oz. from 66.80% to 69.38% from 2013 to 2014
- Aluminum scrap was reduced from 2.92% to 2.15% for 12 oz. and 4.44% to 3.99% for 24 oz. from 2013 to 2014
- Total conversion costs were reduced 12.89% for 12-oz. line and 10.55% for 24-oz. line between 2013 and 2014
- $1,036,640 in cost-saving projects in 2012.
- $1,507,630 in cost-saving projects in 2013 and 2014
- Direct materials usage improved from 2012 to 2014. Varnish usage was reduced by 32%. Ink usage was reduced by 21%. Inside spray usage was reduced by 16%.

For more information contact:
Jon Alder
Director, Group Lean Enterprise
Direct Line +44 (0)20 7227 4197
Mobile +44 (0)7786 197474
Email: jon.alder@rexam.com

Abbott
A Promise for Life

Abbott Diagnostics Longford

Corporate Overview

Abbott is a global, broad-based healthcare company devoted to discovering new medicines, new technologies and new ways to manage health. Abbott products span the continuum of care, from nutritional products and laboratory diagnostics through medical devices and pharmaceutical therapies. Abbott's comprehensive line of products encircles life itself – addressing important health needs from infancy to the golden years. Abbott serves customers in more than 150 countries and has approximately 69,000 employees.

Abbott Diagnostics Longford

Abbott Longford's manufacturing facility was established in 2004 on a 20-acre green-field site on the outskirts of town. The 135,000-square feet building was specifically designed to allow the most efficient flow of product through the value stream from raw material receipt to finish goods shipment. Manufacturing commenced in March 2005 and the first product was shipped in December 2005. There are over 250 employees on-site designing, developing and manufacturing in-vitro diagnostic products.

Products

Longford manufactures 17 different diagnostic reagents as well as calibrators and controls. The product portfolio includes diagnostic kits for thyroid function, fertility & pregnancy, cardiology, renal, metabolic and therapeutic drug monitoring. Products are manufactured to the highest standards of quality and regulated by bodies such as the U.S. Food & Drug Administration and the International Organization for Standardization (ISO). The site's quality policy is *"To improve healthcare by providing high quality, safe and effective diagnostic products."*

Site Vision, Strategy and Behaviors

The site's vision of *"World Class Performance in Everything We Do"* has inspired a site strategy which is based on value propositions of customer loyalty and operational excellence.

The strategy is communicated throughout the organization using site and departmental strategy map/ balanced scorecards.

The continuous improvement culture is supported by behavioral standards which were developed by employees and are underpinned by the guiding principle of *"Treat others as you would like to be treated."* These are intrinsically linked to core employee competencies of Build, Innovate, Anticipate, Set Vision & Strategy and Deliver Results.

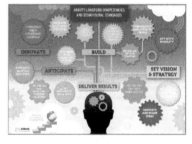

Furthermore, specific Longford Leader Behaviors have been developed to support each core competency. These behaviors and strategic goals are formally assessed, with constructive feedback given to employees, through the talent management system and performance management process.

Continuous Improvement

The Lean management system and formal management walks sustain the continuous improvement culture and incorporate the key elements of leader standard work, visual management, discipline and accountability.

The facility practices daily root-cause analysis to close the gap between expected and actual performance. Value stream mapping, gemba walks, daily kaizen, continuous improvement huddles and waste Identification processes provide inputs for cost & quality improvements which are managed through the product portfolio management process. Through the latter process, >$22 million cost improvements have been implemented and sustained. Since 2006, the site has seen a 576% growth in test volume **and** reduced cost per test by 60%.

They engage in regular kaizen events and execute Lean six sigma projects. The quality system corrective action/preventative action process is based on the DMAIC methodology and all investigators are trained in structured problem solving. Since 2007, they reduced the non-conformance rate by 77%.

The site has completed a formal lead time reduction program, *"Accelerate,"* which led to a 38% manufacturing lead time reduction and decreased inventory holding by 10%. The program had greater than 100 active participants from across all site functions. Furthermore, they have improved planning and control (P&C) processes through the introduction of Class A P&C. The site has invested significantly in automation to improve product quality, productivity and to better utilize the site's highly skilled employees.

Customers

The organization supports new product introduction, customer-focused product improvements and provides specialized customer support. Through proactive customer engagement, they seek to understand and deliver what the customer wants. We interact and support customers in their own laboratories and the site has hosted multiple customer visits and forums.

People & Environment

The health and safety of employees and the protection of the environment is paramount. The facility won the Abbott Global Environment Health & Safety & Energy Plant of the Year Award in 2009, 2011, 2013 and 2014 and was runner–up in 2007, 2008 and 2010.

Abbott Longford is a zero landfill site and is certified in Environmental Management System ISO 14001 and Occupational Health and Safety Management Series 18001. They currently have 3,258 (30 June 2015) lost time accident free days. They also have had a 23% decrease in energy usage in the last four years while increasing production output by 68%.

The Corporate Social Responsibility Initiatives have won multiple local and national awards with employees volunteering 2,126 hours in 2014. Several employee-focused strategic initiatives include communication forums, a Live Life Well & Work Life Balance Program and employee centric reward & recognition systems.

The facility encourages employee career development through growth planning, internal education and skills development programs, cross-training and job rotations.

For Further Information Contact:
Seán Kelly, Business Excellence Manager
Abbott Diagnostics, Longford, Ireland
Tel: +353 43 3331090   **Email:** sean.s.kelly@abbott.com

ECA
GUATEMALA

Beverage Can North America Joint Venture – Guatemala Can Plant

Corporate Overview

Envases de Centroamérica, S.A (ECA) was formed as a joint venture in 2007 by the partners Envases Universales and Rexam.

Envases Universales is headquartered in Mexico City, Mexico. It is comprised of 45 plants in 7 countries. The industrial group has three divisions Food/General line, Plastics (PET), and Aluminum.

Rexam is a leading global beverage can maker with sales in the region of $6.5 billion in 2013. They have 55 can making plants in more than 20 countries and employ around 8,000 people. Headquartered in London, England.

ECA has four embedded core values: Teamwork (trabajo en equipo), Continuous Improvement (mejora continua), Loyalty (lealtad) and Commitment (comprosmiso). Teamwork to get the best outcome; Continuous Improvement that is always seeking perfection; Loyalty to each other in a respectful way; and Commitment to customers, the community and environment. ECA has been committed to a path of Lean enterprise across all operations as a way of putting values into practice.

The Guatemala Facility

The Guatemala Can Plant, based in Amatitlán, Guatemala, started operation in 2006, by Envases Universales. In 2007, the joint venture took place between Envases Universales and Rexam, and Envases de Centroamérica, S.A. was born.

Of the 122 people on-site, there are 98 involved in plant operations and management, and 24 people are responsible for finance, logistics and sales. The 163,600-sq. ft. facility has one aluminum can manufacturing line that is fully dedicated to customers such as Coca-Cola, InBev, SAB Miller and Pepsi, with a total production of 4.2 million cans a day. The plant operates 24 hours a day, seven days a week, and is driven to exceed customer expectations.

Product and Process

The site is capable of producing one size of beverage can, 12 oz., for 26 customers, resulting in over 440 active label options. The ability to continuously improve changeover times to meet requirements for smaller lot size requirements and make faster deliveries is at the heart of maintaining a competitive advantage.

To make a beverage can, the key process steps involve stamping and drawing aluminum discs from rolled stock to form a can shape. The cans are then washed before being labelled using in-line screen printers. They are then shipped to customers' filling locations across Central America and the Caribbean.

Speed and quality are the key drivers of process performance and SuperCEP (Statistical Quality System), manufacturing operations management and enterprise manufacturing intelligence provide real-time production data and process monitoring capability.

Continuous Improvement Process

Rexam became a partner in 2007, and after start-up, one of the company's major contributions was the Lean enterprise culture. Rexam has operated a global system of Lean enterprise since 2003, structured around building capability in specific toolsets of culture, VSM, 5S, TPM, SMED and Six Sigma.

Rexam's best plants were invited to participate in a more sophisticated assessment known as "Beyond Gold," where the implementation of lean systems is appraised. A system of merit is used to recognize achievement starting at *Emerald*, then *Sapphire*, and finally *Diamond*. In recognition of their Lean efforts, the first Rexam Lean enterprise audit was in 2012. The Guatemala plant applied to Beyond Gold assessment and achieved Sapphire level, and subsequently moved to Diamond in 2013.

Improvements are identified and delivered using a matrix organization structure with functional floor leadership called "The Grill" (initiated in 2007). Between 2012 and 2014, 96 kaizen events took place with 100% participation from the people of the plant.

Guatemala Achievements

- 2.8 million hours without a lost time accident (LTA)
- Benchmarked against North and South America Rexam plants, ranked top three in six categories and best-in-class in ink, gas and hydraulic oil usage (April, 2014)
- Lean enterprise achievements:
  -Beyond Gold *Sapphire* in 2012
  -Beyond Gold *Diamond* in 2013
- Recognition of decoration excellence from LatinCan group and *Canmaker* magazine, 2008-2014
- The plant's OEE performance improved from 80.3% in 2008, to 94.0% in 2013
- Aluminum scrap has reduced from 3.81% in 2008, to 1.71% in 2013, for a total reduction of 55%
- FSSC 22000: 2013 Food Safety accreditation

Safety & Environmental

- Zero lost time accidents (LTA) for five consecutive years from 2010 to 2014
- Rexam Chief Executive Officer Safety Award for achieving one million hours with zero lost time incidents in 2011
- LTA rate has reduced steadily from 1.05 in 2009, to 0.19 in 2013
- Total Incident Rate has reduced steadily from 11.19 in 2009, to 3.49 in 2013
- Evacuation drill time improved by 71% in response time, between 2011 and 2013. Guatemala is part of the "Ring of Fire," an area where a large number of earthquakes and volcanic eruptions occur in the basin of the Pacific Ocean
- Promoted the "Infinitely Recyclable" message of the aluminum can in Guatemala with billboards and educational programs
- Electricity and LP Gas consumption reduced by 30% between 2008 and 2013

Rexam PLC
4 Millbank
London SW1P 3XR
United Kingdom

ECA
GUATEMALA
Beverage Can North America Joint Venture – Guatemala Can Plant

Quality
- Seven certified black Belts (quality manager, engineering manager, production manager, industrial safety and environment, continuous improvement leader, SQC analyst and projects coordinator)
- Eight certified green belts (quality process analyst, predictive maintenance technician, front end responsible, back end responsible, customer service, general accounting, chemical area coordinator and raw material analyst)
- 107 yellow belts (quality, engineering, production, finance, sales and logistics, IT and human resources areas)
- Number of customer complaints per billion produced has reduced steadily from 112.07 in 2007, to 8.8 in 2013, for a reduction of 92% over those years
- Coca-Cola, InBev, SAB Miller and Pepsi Certified
- Real-time quality system management and statistical process control SuperCEP (Statistical Quality System).
- Real time visual plant manufacturing system and downtime management (ActivPlant)
- Automated in-line vision systems to detect mix label and flange-necker defects
- Support customers on-site in order to improve their can line run-ability
- FSSC 22000: 2013 Food Safety accreditation

Employee Morale
- Absenteeism has remained at low levels during the last seven years
- Turnover was reduced from 2.23% in 2008 to 0.64% in 2013
- 95% of the people are extremely satisfied with ECA as a place to work and 97% of the people are proud to work for ECA
- Offers English speaking classes because the technology of can making is English language-based. This is an advantage for technical learning
- Six health fairs are offered annually with an in-plant medical doctor

Delivery performance and cost reductions
- OEE has increased from 80.3% in 2008, to 94.0% in 2013
- Installed capacity has increased from 73.3% in 2008, to 91.5% in 2013. This resulted in additional 575 million cans produced annually
- Aluminum scrap was reduced from 3.81% in 2008, to 1.71% in 2013
- Total conversion costs were reduced 18.7% between 2008 and 2013
- $992,000 in Lean Enterprise/Six Sigma cost saving in 2012 and 2013
- Direct materials usage improved from 2008 to 2013. Ink usage was reduced by 22%

People
ECA believes its people are its most important asset. The ECA team started with no can-making experience in the country of Guatemala. Ninety-seven percent of the people working at the plant are Guatemalan and developed expertise through a strong commitment to learning and improvement. ECA is proud to retain and develop 70 team members (122 currently) since the plant startup.

The location is used as a training site for methods and processes with our joint venture partners, Rexam plants from the United States, South America and Europe, as well as Envases Universales from Mexico. They enjoy having these people visit the plant because they always learn new things from them.

For more information contact:
Jon Alder
Director, Group Lean Enterprise
Direct Line: +44 (0)20 7227 4197
Mobile : +44 (0)7786 197474
Email: jon.alder@rexam.com

Clonmel, Ireland

Corporate Overview

Abbott is a global healthcare company devoted to improving life through the development of products and technologies that span the breadth of healthcare. Abbott serves patients in more than 150 countries and employs approximately 70,000 people. The company, headquartered in Chicago, acquired Guidant's Vascular business in 2006, which became part of the Abbott Vascular Division, to create one of the world's leading medical device companies.

Abbott Vascular Clonmel

Clonmel is one of Abbott Vascular Division's 9 sites, it has been in existence since 1998 (initially as Guidant, and then in 2006 as Abbott). The vision is "*to improve the lives of patients and their families around the world through our passionate commitment to innovation in cardiac & vascular science.*" This is driven by divisional strategic themes, which waterfall through to everyone:

- Product Availability
- Quality & Compliance
- People
- Simplify

Product Availability;

The site's commitment to customers is to have the right products, in the right place, at the right time. As a manufacturing organization, it is the site's responsibility to make sure the supply chain fulfills that commitment.

Quality and Compliance;

As part of the medical device industry, the facility operates in a highly regulated environment. To keep this commitment in the fore-front of employees, the site's quality policy, "*Built as is intended for my Family*" is clear, and translates easily to all employees. As a result, employees act with courage and integrity to exceed customer expectations.

People:

At the heart of the site's strategy is people. They are committed to putting people first, promoting safety and respect, recognizing the ability and diversity of the staff. They strive to have all employees proud to work in a safe environment where they are empowered to improve their roles and the site.

Simplify & Focus:

Manufacturing technical products in a regulated environment provides a competitive challenge. It is important to remain agile and get products to market first, and having robust systems is key to this. Through continuous improvement, application of lean and the appropriate use of tools; the site is committed daily to meeting this challenge.

Gross Margin:

In an ever-increasing competitive market, Clonmel adds value to the division's margin by continually reducing manufacturing costs. Every dollar saved in manufacturing costs significantly impacts the distribution margin. The site drives cost competitiveness through all processes.

All projects and initiatives are filtered through these themes ensuring full strategic alignment, where the belief exists that success is twinned with an unstinting commitment to operational excellence. Within this, the site is both positioned and recognized as the divisional "Centre of Excellence" for DES (the Eluting Stent) and Implant (Bare Stent) manufacturing.

The site has 1200 employees primarily dedicated to manufacturing but also includes process development. The 252,000- sq.ft. facility sits on 19 acres has 3 distinct areas manufacturing products for the treatment of coronary and endovascular diseases.

- DES (Drug Eluting Stent manufacturing)
- Implants (Stent manufacturing)
- SDS (Stent Delivery Systems or Catheter Manufacturing)

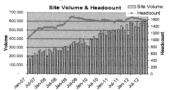

Site volume increased by over 300% since 2007, yet headcount has remained steady since Jan 2009

Product

Abbott Vascular provides customers with products such as DES Xience V and Xience Prime, which are the world's leading coronary stent systems. This stent when placed into narrowed, diseased arteries slowly release a drug, and prevents further blockages to the stented artery. The site also manufactures its predecessor bare metal stent products (non-drug coated). Both products include the stent (drug coated or bare) and a catheter delivery system needed to place the stents in situ during the patient's angioplasty procedure. Both of these sub-assemblies are also manufactured at the site.

The DES finished good volumes can be seen below, which have continued to grow despite difficult market conditions to an extent where Abbott Vascular is now the market leader. Due to the site's excellent performance, Clonmel now provides 100% of Abbott's DES finished goods products and stents.

Due to the nature of the products, quality is paramount. The quality system has ensured they put patient's first while achieving dramatic continuous improvement in the site's performance in achieving double-digit reductions in non-conformances while also achieving double-digit reductions in costs.

Clonmel, Ireland

Policy Deployment

The policy deployment system ensures line of sight for all by visually creating the links between the five-year divisional strategy and all employees. This is enhanced by the site strategy and performance excellence system and is supported by daily meetings, visual communication boards and monitoring systems.

Through alignment with the divisional strategy, the site developed strategic programs to reduce product cost by 50%. This came about through confidential benchmarking and drove the DES unit cost reduction of 73% and productivity improvements of >300%. This is aligned with the vision for Clonmel to be the division's *Site of Choice*.

The performance excellence system manages goal-setting for the site and employees and are linked through the talent management software system (TMS) which cascades goals from strategy to the manufacturing floor, through managers to associates. Also, the policy deployment process combines the divisional requirements and the site's input of what needs to be achieved into an aligned set of priorities. The quarterly prioritization system ensures the effort to achieve these goals remains focused.

Quality System

The Abbott quality system ensures there is clear line of sight from the governing policy at the divisional level, all the way to the daily line meetings and work instructions. It ensures all issues that occur, whether isolated events or trended anomalies, are investigated, root-caused and improved upon. If these issues arise from the manufacturing floor through a non-conformance or as a customer complaint, it is dealt with appropriately as part of the Quality Reporting System (Left).

The CAPA (Corrective and Preventive Action) system has strict guidelines ensuring problem-solving tools are used to conduct the investigation, e.g. FMEA, 5Ms, Ishikawa, Contradiction Matrix. All documentation to support systems is aligned, ensuring constancy of application.

Continuous Process Improvement

The site in Clonmel has been on a lean journey since 2004. Throughout that time, tools, systems and the culture have developed to a stage where the maturity of the site encourages continuous improvement, through A3/kaizen DMAIC, 5S, VSM, SMED, standard work and six sigma.

The lean program started with the application of lean principles in the operations areas through a 14-step lean process. This is a transformational lean change process that teams commit to for 4-5 months. The first project began in 2005 and this process is still used today. The last two 14-step projects led to 38% and 37% productivity improvements balanced with a 50% quality improvement.

In 2007, the facility commenced a local process improvement (LPI) system to harness the continuous improvement (CI) ideas of all employees. They developed this system in 2012 through CI huddles, where every associate is engaged and responsible for localized problem solving. This huddle process came about from benchmarking Shingo recipient organizations; they are one of the key methods to address employee suggestions. With over 4400 process improvement or safety suggestions implemented in 2012 (vs. 2000 in 2008), the site continues to make progress toward operational excellence. The six sigma program started to accelerate in 2008 when they set a 5-year goal to reach 80% of the technical staff receive green belt and black belt certification (approx. 100 people), ensuring they use the tools as part of their daily work. They tackle key business improvements and knowledge development to support the drive to remain "Centres of Excellence." The facility achieved this goal in 2012, and in 2012, the Lean and six sigma programs led to cost improvement savings of $11.5M (vs. $1M in 2007).

In 2008, the Clonmel site began a Class A journey and in 2009, they received Oliver Wight Class A accreditation for Planning and Controls. This is awarded for business excellence systems related to customer service, in particular the implementation of the Integrated Business Model (also known as Sales & Operations planning or S&OP) and monthly business review systems. The latter is to review the key performance indicators. The combination of the two keeps the business and operating plans aligned as well as focusing efforts on strategic planning. The site reapplied to Oliver Wight for accreditation in 2011 and was again successful.

Within Abbott Vascular, the Clonmel site has led the way in terms of building relationships with suppliers, resolving issues and driving improvements, including training them in six sigma and quality systems. They review monthly and quarterly supplier performance in a formal business review. In 2008, the site began holding an annual Supplier Appreciation Day which drives partnerships with the highest performing key suppliers, driving CI in the supply channels. This has aided material cost savings of $10.8M or ~7% per year Purchase Price Variance or PPV in the last 5 years. This is above the industry norm of 4-5%, and allows process engineers to build relationships with those supplying their parts.

Conclusion

Clonmel is considered one of Abbott's flagship manufacturing facilities. The site hosts many benchmarking sessions within Abbott to share what they have achieved and to learn from others along their journey toward operational excellence.

For more information please contact:

Pat Kealy – BEX Manager,
Tel; +353-52-6173110, Email; pat.kealy@av.abbott.com

"Built as if Intended for my Family"

Overview

Barnes Aerospace Fabrications, Ogden Site (BAFO) is located in Ogden, Utah. Barnes Aerospace focuses on two all-encompassing strategic platforms: 1) **Our Strategy** - Providing a product life-cycle solution for customers; and 2) continual pursuit of **True North** – Improving the lives of employees, customers, shareholders, and communities. BAFO provides superior manufacturing solutions to the world's major turbine manufacturers, commercial airlines and military original equipment manufacturers. As a business of Barnes Group Inc., BAFO takes advantage of the financial stability and resources of a large company, while offering the quality and service associated with smaller businesses.

BAFO's success stems from the enthusiasm of associates as they strive for excellence. The lean journey has transformed the culture from a tools- and systems-driven approach to an employee empowered culture which embraces the following values:

1. Continuous Improvement
2. Open mindedness and communication
3. A learning organization focus
4. Total organizational team work
5. Empowerment

These values, paired with the drive of associates, have allowed BAFO to continue achieving significant sustainable improvements year over year, making the True North a part of the organization's DNA.

Throughout the lean journey, BAFO has continued to build on accomplishments, reducing waste and increasing efficiency while growing sales revenues by over 450%.

Achievements

Environmental Health and Safety
- ISO14001 certified
- One million hours worked without a lost-time accident achieved in 2010 for the third time in BAFO history
- Currently over 650,000 hours worked without a lost time accident
- Behavior Based Safety Program

Quality/Cost/Productivity
- 53% reduction in scrap and rework as a percent of sales
- 64% reduction in internal escapes
- 55% lead time reduction in days to manufacture
- 23% decrease in inventory days on hand
- 65% decrease in freight costs per pound

Customer Satisfaction/Delivery
- 28% increase in on-time delivery
- New product introduction on time delivery increased from 74% to 100%

Community Service
- Annual United Way campaign and many other charitable contributions
- Tutoring at a local elementary school
- Year-round community outreach program and quarterly sponsored events for underprivileged children
- Barnes Aerospace Engineering Scholarship worth $5K, Weber State University
- Service in community development and educational boards

Awards and Recognition
- Utah Manufacturing Association and Utah Labor Commission Workplace Safety Award for Outstanding Safety (2012)
- Shingo Silver Medallion 2011
- Utah's Best Places to Work, runner-up in 2009/2010/2012
- Utah Manufacturing Association "Manufacturer of the Year" runner up (2011)
- Bravo Award; Corporate Sponsor of the Year, United Way of Utah (2010)
- Manufacturer of the Year Award for C. I. (2009), Utah MEP
- Premier Supplier Award, Bell Helicopter (2007)
- Platinum Supplier, Northrop Grumman

People

BAFO associates have a long history of creating a culture that strives for continuous improvement and customer satisfaction in a safe working environment. Associates make the True North strategy a success by 1) working as a cohesive team; 2) continual pursuit of customer satisfaction through on-time-delivery, uncompromised quality, and pro-active communications; 3) profit growth for shareholders; and, 4) volunteer community service.

They are benefiting from a 230% increase from 2007 in documented and implemented improvement suggestions - excluding the countless employee "just do it" daily Kaizen implemented improvements - which are leveraged across the organization.

Process

The lean journey began in 2000 by focusing on the implementation of systems and tools from the Toyota Production System. From 2000 to late 2006, improvement activities were driven from single event-based activities, pushed down to associates from the leadership team, yielding low levels of sustainment.

In 2007, Ogden's leadership team focused on teaching, empowering, and aligning systems, tools and behaviors to achieve maximum results. By unleashing the untapped human creativity of associates, they are experiencing step change and sustained improvement activities across the organization.

The leadership team performs structured daily GEMBA walks, ensuring that the entire facility is engaged on a weekly basis. Leadership serves to support and remove constraints in order for associates to deliver value and drive continuous improvement. Associates review performance measurements, as well as showcase their team's newly implemented ideas.

Product

BAFO specializes in the fabrication and assembly of high temperature alloy, complex airframe and engine components with particular expertise in superplastic, hot and cold forming, welding, milling and cutting technology. Working in compliance with AS9100C standards and seven (7) special process NADCAP certifications, BAFO delivers the highest quality and best value solutions to customers.

Facility

In 2009, a 160,000-square-foot manufacturing facility was built using the most current lean flow tools. All associates participated in at least one Kaizen event centered on flow, in order to develop cell and support area layouts. The facility houses three value stream operating units, each of which is focused on new product introduction, sheet metal fabrications, and airframe assemblies.

Corporation

Barnes Aerospace Fabrications Ogden site is an operating segment of **Barnes Group Inc.** (NYSE:B), a diversified global manufacturer and logistical services company with 60 locations on four continents and 4800 employees delivering on our promises to our customers, ensuring exacting performance, superior support and service, and impactful results.

For more information please contact:
Steve Moore, General Manager
Barnes Aerospace Fabrications
1025 South Depot Drive
Ogden, Utah 84404-1390
Phone: (801) 917-2001
Mobile: (801) 388-1734
Fax: (801) 917-3000
Email: smoore@barnesaero.com

OVERVIEW

DePuySynthes is a subsidiary of Johnson & Johnson, one of the world's largest and most diverse healthcare corporations. They offer a broad portfolio of orthopaedic and neuro products for joint reconstruction, trauma, spine, sports medicine, neurological, craniomaxillofacial, power tools and biomaterials.

The DePuySynthes plant, based in Cork, Ireland, is part of the Joint Reconstruction business and currently manufactures approximately 40% of worldwide volumes. The plant was established in 1997 and the volume has steadily increased from 100k units in 1998 to 1.6 million units in 2012, with a plan to ship over 2 million units in 2013.

PRODUCTS

The Cork site predominantly manufactures hip, knee & shoulder replacements. There are approximately 1700 SKUs (Stock Keeping Unit) manufactured at the facility.

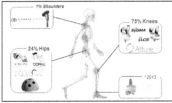

PROCESS

The product starts as a raw Cobalt Chrome alloy billet and is taken through a series of technologies transforming the product into a high-precision device which is cleaned and packaged on-site. The processes are organized into value streams based on the product type. There are five main value streams on-site, the Foundry, Femorals, Trays, Poly and Hips value streams.

CONTINUOUS IMPROVEMENT JOURNEY

In 2003, the company began to introduce lean tools such as CONWIP and kanban systems, overall equipment effectiveness (OEE), value stream mapping and supplier integration. This led to some significant lead time reduction and cost improvements, but there was acknowledgement that this tool-based approach was not sustainable or culturally ingrained.

In 2005, there was an unexpected increase in demand which the plant was not in a position to respond. The market norms had shifted considerably, driving the need for a more responsive and flexible manufacturing facility. To succeed in this new environment the site needed to build a high-performance culture based on continuous improvement. In 2006, the plant began a cultural, physical and organizational transformation. The site created a vision for 2010 based on doubling capacity, while maintaining the same headcount and floor space. The transformational map was constructed under four work streams, a lean program, change management, new product introductions, and compliance excellence. Below is a snapshot taken in 2009 of the progress towards that goal.

In 2007, machines were physically moved from grouped processes to a value stream (VS) layout. This was completed to promote flow and enable pull within each area. This involved the relocation of over 400 pieces of equipment.

During this transitional phase the organization structure was changed from functionally based to a value stream structure. This meant the creation of a VS manager with overall responsibility for the VS and a support team with shared objectives. The support teams were then relocated into custom made 'pods' (pods are office units for manufacturing support staff located within the VS). The space for these pods was created through the space saved through the lean layout. As the VS structure progressed, this led to aligned goals and objectives (G&Os) and a profit & loss account by value stream which enabled better decision making.

All decisions within Johnson & Johnson are based on the organization's Credo. It is a set of values detailing responsibilities to customers, patients, doctors and nurses that use their products; to employees; to the communities in which they work; and finally, to stockholders. In 2008, the site saw the need to create a highly engaged and adaptable workforce to meet the ever-changing business demands. This change in culture was enabled through the creation of behavioral standards. These are five statements that govern the commitment around how they work with other. Each person on-site contributed to their creation in the process shown below.

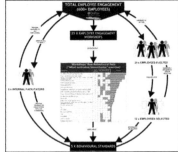

The company is driven by its vision, the true north for continuous improvement. From a maturity perspective, they are now in the process of developing a third site vision, i.e. the 2020 vision.

Standard work was piloted in one value stream in 2010. It has now been rolled out across the manufacturing site. All documents for new products and product transfers are now written in the new standard work format. Each document is structured around major steps, key points and reasons why the step is there. The standard work roll out is underpinned by Training Within Industry (TWI) training. All TWI trainers are trained by certified coaches.

The use of value stream mapping (VSMs) as a system dates back to 2004. As the facility has matured, they have now integrated the VSMs into the CII process (Continuous Improvement through Innovation). This process creates a scheduled pipeline of projects that leads towards the 'target' process state, creates the 'business plan' for the forthcoming year and creates a list of feasibility projects that can improve the quality or cost of products. This process has also been used to map the extended supply chain, from tier-two raw material suppliers, through to primary and secondary distribution centers. The Integrated Management System (IMS, introduced in 2011) is an overarching system that governs how they run the value streams. It entails a bottoms-up hourly management process, through to a weekly senior management gemba review. The IMS comprises the following sub-systems; Kaizen, WIP control (FIFO, heijunka, and ticket systems), Theory of Constraints, Problem Solving, Capacity and Headcount Model.

In line with the Credo values, making the customer central to all they do is paramount. The organization runs a customer connection program on-site. This consists of talks at quarterly communications from patients (normally relations of staff) who have benefited from certain products. They also schedule visits to live surgeries for employees, as well as bring surgeons and theater nurses on-site to view manufacturing processes.

The company also has a responsibility to the community they work in, and they run a strong community outreach program. In the last 15 years, they have contributed €750k and 8000 hours of employees' time to the community and charity work. These charities are chosen annually by employees.

RESULTS HIGHLIGHTS:

2006-2012 Reduction in lead time from 16 days to 4.5 days

2006-2012 Doubling of capacity within the same footprint

2007-2013 Increase in productivity of 130% for the site (average annual productivity gains of 20% + per annum)

2009-2013 10%+ year-on-year cost improvements as a percentage of manufacturing cost

2008-Present 85%+ engagement levels based on the J&J Credo survey

100% of consumables and **40%** of raw material under Vendor Managed Inventory (VMI) control, resulting in elimination of on-site warehouse in 2009

100% on-time delivery of new product launches, in 2014 40% of volume will have been launched post 2012

AWARDS/RECOGNITION:

2009: The site was certified to ISEN 16001 Energy Management Systems (first site in Europe)

2009: Engineers Ireland, Continuous Professional Development (CPD), Company of the Year award

2010: Awarded the overall winner in the IMDA (Irish Medical Device Association) for 'Manufacturing and Operational Excellence' as a result of the ongoing Business Transformation and Cultural program at the Cork facility

2011: First company in Europe to become ISO 50001 (Energy) certified

2012: Worldwide J&J Sustainability Energy Excellence Award

2013: Worldwide J&J EH&S excellence award focused on Metal Usage reduction

2013: IITD National Training Award, for the application of Training Within Industry (TWI) training

For more information contact:
Joe Healy, Operational Excellence Manager
Direct: +353(21)4914339
E-mail: jhealy1@its.jnj.com

James Winters, Plant Manager
Direct: +353(21)4914647
E-mail: jwinter9@its.jnj.com

Newsprinters (Eurocentral)
Operation Overview

Corporate

Newsprinters are the printing division of NewsUK (formerly News International) and are a wholly-owned subsidiary of News Corporation (the largest news and information service provider in the world). With three purpose-built facilities located throughout the United Kingdom (Glasgow, Liverpool and London) which house the very latest state-of-the-art printing technology and equipment that ensures Newsprinters remains at the forefront of the newspaper printing industry.

In addition to the economic crisis of the past six years the Newspaper Industry has been faced with many challenges; dropping readership (Fig 1), declining advertising revenue, increased supplier and logistics costs and competition from online and electronic media.

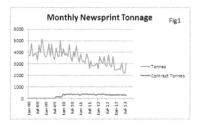

Monthly Newsprint Tonnage Fig1

Against the background of this global phenomenon, Newsprinters has set the vision **"To be recognized as the best Newspaper Manufacturer in the World"** and in 2007 set off on a continuous improvement journey to achieve it, thus sending a clear message that they aim to be the best in all areas of business, whether it be meeting the customer expectations of quality products on-time in-full, health and safety, environmental performance or staff engagement.

The four aims of the business are:

- To be the most efficient and cost effective newspaper manufacturer.
- To be the newspaper printer of choice.
- The culture will be supportive, inclusive and will encourage high performance and innovation.
- To practice industry-leading standards of health & safety compliance and environmental performance.

Eurocentral Site, products and processes

Newsprinters – Eurocentral (Scotland) was officially opened in April 2008 by Scotland's First Minister, the Rt. Hon. Alex Salmond and News Corporation chairman and chief executive Rupert Murdoch. This £56 million investment in Scotland

also has the distinction of being the home to the largest and fastest newspaper printing press in the world.

As Scotland's largest newspaper manufacturer, they currently print approximately three million newspapers per week; *The Sun, Sunday Sun, The Times & The Sunday Times*; additionally, they contract print *The Daily Telegraph, The Sunday Telegraph, Glasgow Solicitors Property Guide, The Edinburgh Evening News* and others making them the largest contract printer in Scotland. The site employs 120 people: 60 in operations, 14 in management, 16 in engineering and 30 in facility operation. The plant was designed with flow principles ensuring minimum distances between key processes.

Newsprinters uses the lithographic printing process to produce high-quality color newspapers in tabloid or broadsheet format which are then packaged for on-going distribution throughout Scotland. The cycle time is approximately 10 minutes and the output of both presses is a maximum of 172,000 copies per hour. The quality and output is monitored both automatically and manually at every stage of the process and this data is stored in bespoke management systems for analysis of long-term trends and issues.

Continuous Improvement

Newprinters's continuous improvement journey started in 2007 (Fig2) when they implemented the first asset champion program and 5S pilot project, while at the same time introducing standard work processes. Since this time, they have successfully achieved three certified ISO standards: ISO9001:2008, ISO14001:2004 and BS OHSAS18001:2007.

NCIS - Road Map Fig2

The implementation model of ISO systems was based on cross-functional teams (operator/technical/managers) and has proved to be very successful in delivering excellent business results and developing the "one" team culture and behaviors.

To achieve these ISO standards, Newsprinters divided the business into 12 Quality Process (QP). These vertical areas cover the full scope of the site's activities, and when one looks horizontally across the areas one sees the complete value streams (information and product flow). And, it is this QP model that they have built all improvements on, with each area creating integrated teams and having the autonomy of a separate business to make decisions and improvements based on their internal and external customer's needs.

Following the successful implementation and certification of the three ISO standards, they developed the Newsprinters Continuous Improvement System (NCIS), built on a foundation of Integrated Teamwork and Communications, supported

Fig3

by three main pillars; Organization, Standardization and Engagement, topped by Leadership (Fig 3). This system has six inter-related sub-systems:

1. **EA** = Enterprise Alignment (Policy Deployment System).
2. **TPM** = Total Productive Manufacturing (in house Continuous Improvement System).
3. **IBMS** = Integrated Business Management System, integrating 9001, 14001 and 18001.
4. **T&D** = Training and Development.
5. **NEG** = Newsprinters Environmental Group System.
6. **Asset Champs** = Staff and managers designated/assigned asset responsibility.

The Total Productive Manufacturing (TPM) System is a holistic continuous improvement system designed to improve operational efficiency and effectiveness by: eliminating or reducing all safety losses, waste, breakdowns and improving our overall performance. TPM is rolled out through a series of workshops that are scheduled over a one-year period and teaches integrated teams the fundamental quality principles as taught by Dr. Deming and the use of Lean tools as developed by Taiichi Ohno and Shigeo Shingo (TPS). The knowledge and skills developed through this system allows these integrated teams to map their process, solve complex problems, reduce waste and identify opportunities for improvement.

Newsprinters ensures line of sight and visibility of corporate strategy and objectives through all levels of the business by using Policy Deployment and align the enterprise with business communication and management systems. This system guarantees the alignment between communications, decision-making processes and employee engagement through all levels of the business and ensures that policy deployment is directly linked to all QP areas and to each member of the staff, who all have their own personal policy plans which again align both their area objectives and their personal development objectives.

Achievements

Lean Enterprise
- 5 BITs diploma and 1 BITs level 4 degree
- 80% Staff trained in Newsprinters TPM (4500hrs)
- SMED improvements to press change overs leading to new contract awards
- 300+ Improvement activities
- £1.0million savings to the business

- TPM improvements leading to greater equipment reliability
- 15 SMED activities on engineering single point of failures

Safety & Environmental
- ISO 14001 environmental accreditation
- OHSAS 18001 safety accreditation
- Creation of Bio-diversity garden + School partnership
- UK Newspaper Awards - Environmental Initiative of the Year - 2011
- UK Security Excellence Awards - Environmental Initiative of the Year – 2011
- BPIF Vision in Print – "Environmental Practice" Audit No. 1 in Industry - 2010
- Zero lost time accidents since 2009
- 30 trained first aiders
- Zero RIDDOR reports in since 2007
- Accreditation to deliver NEBOSH Safety Training

Quality
- ISO 9001 quality accreditation
- UK CSSA Cleaning Industry – Cleanest Manufacturing Site in the UK - 2011
- UK Newspaper Awards – Printer of the Year 2013
- BPIF Vision in Print – "Best Practice" Audit No. 1 in Industry – 2010

Morale
- 7 x "Big Sort" employee-led 5S programs
- 66% People score in employee engagement survey (UK average – 56%)
- Staff survey results show 90% approval (Good to excellent) for site communications
- Cross-functional team ownership of all ISO and CI/TPM initiatives

Performance Delivery & Cost Reductions
- Plant efficiency up by 30% since 2007
- Newsprint waste reduced by 40% since 2007
- Gas consumption reduced by 300% since 2007
- Electricity consumption reduced by 30% since 2007
- 0.5% technical downtime average 2013

For more information contact:
George Donaldson
Group Continuous Improvement Manager
Direct Line: +44(0)1412029389
Mobile: +44(0)7919470861
Email: george.donaldson@news.co.uk
Newsprinters (Eurocentral) Ltd
Byramsmuir Road
Holytown ML1 4WH
Scotland
United Kingdom

ETHICON INC.
Juarez, Mexico

OVERVIEW

Ethicon has been the world leader in manufacturing of surgical sutures for over a century with a total share of over 70% of the global market. The Ethicon Juarez facility initiated operations in 1999 with an initial staffing of 19 people supporting the manufacturing of a family of products transferred from Australia. From 1999 to date, Ethicon Juarez has been growing in part due to the Johnson and Johnson Asset Reconfiguration strategy, but primarily due to competitive labor cost and competitive productivity. It currently has a headcount of 1,196 people, both wage and salary personnel. Ethicon Juarez supports 20% of the total Ethicon worldwide suture volume.

Ethicon Juarez vision is "To be the best world class manufacturing operation through robust process capability, improving lives with superior customer satisfaction at the most competitive cost".

4,970 SKUs are manufactured in the Ethicon Juarez facility; synthetic absorbable, natural and non-absorbable synthetic are some of the surgical suture families produced; Thermachoice™, a device intended to ablate the endometrial lining of the uterus in premenopausal women with menorrhagia (excessive uterine bleeding) and SecureStrap™, a surgical device for laparoscopic (minimally invasive) hernia repair procedures are in the category of medical devices families.

PROCESS

As part of a global supply chain, ETHICON Juarez resolves issues and executes process improvements beyond the four walls of its site by working with suppliers and customers. The entire supply chain is taken in consideration as part of the plant management strategy development for current and future years.

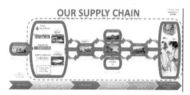

A Production System has been put in place as a sustainable process in the Ethicon Juarez manufacturing plant. This process enables the identification of opportunities for improvement and activities which enhance the product flow and quality of the product that reaches the customers. "Plan, Do, Check and Act" are the four key elements of the system.

Standard Lean Production Line was designed to serve as a catalyst for faster and flawless execution anticipating constraints for any new product transfer or line reengineering. Standard work for the production line operators inclusive of operator balance, product takt time and change-over times were put in place in 80% of the total manufacturing floor. Standard work has also been defined and implemented for 100% of material handlers and supervisors. New product lines such as of SecureStrap™ include standard work from day one of production.

The Plant is divided into two Business Units - Split flow and Specialties.

Split flow Business Unit is composed of 3 lines dedicated to handling raw materials, 7 Lines dedicated to sutures assembly and 1 line for devices (SecureStrap™).

Specialties Business Unit has 6 lines dedicated to sutures assembly, 1 for devices (Thermachoice™) and 2 specialized sutures lines based mostly to meet special orders with low volume or specific requirements (Make to Order).

Sutures: Material is received and stored in the plant's warehouse and it is then moved to a kitting area in the controlled environment manufacturing floor. Material is then assembled, packaged, and taken to the warehouse for shipment to other Ethicon or J&J facilities in the US for further processing and distribution.

Devices: Material is received in the plant's warehouse and then moved to the manufacturing floor where it is assembled, packaged and shipped to the US.

CONTINUOUS IMPROVEMENT

Embed design to value for customers, assuring a safe compliant and organized environment, integrated support systems, developing measures to promote the right behaviors and maximizing asset care are just some of the critical elements Ethicon Juarez continues to practice and reinforce to achieve a superior work environment.

Hoshin Kanri represents the defined process used to communicate and cascade all strategies, goals & objectives to all levels of the organization. This management system is used to ensure the participation of all employees, with corporate goals that cascade down to specific metrics related to the production line. Communication meetings, metrics boards, Gemba walks and electronic dashboards are part of the structure in this Policy Deployment.

Key initiatives ensure productivity improvements and are part of the day to day operations such as "Servant Leadership Culture", "Do it Right, Do it Better", "Development and Recognition", "Continuous Improvement Programs", "Outside the 4 walls" and worldwide J&J "CREDO". The implementation of continuous improvement programs has represented the primary engagement tool that has enabled Ethicon Juarez to achieve its goals in support of productivity improvements. The integration of the Six Sigma and Lean Manufacturing has been the strategic approach used to reach results and growth following the rigor of the DMAIC (Define, measure, Analyze, Improve and Control) and DMADVV (Define, Measure, Analyze, Design, Verify and Validate) methodologies.

Kaizen is now part of Ethicon's culture, managed by four programs that have become the way we solve problems at the Ethicon Juarez site. These programs include our New Idea and Suggestion

Program, Kaizen, Kaizen Blitz and Kaizen Green. These programs lead by our production associates, were implemented to manage the associate's ideas and continuous improvement.

Even though Ethicon Juarez Plant is the youngest of the Ethicon's facilities, it is considered a benchmark facility in lean manufacturing processes, floor visual management, communication flow and most importantly...people engagement.

ACHIEVEMENTS

Improvements in Quality, Productivity and customer service have been achieved each time processes are transferred from other sites. As an example Thermachoice™ complaints rate was reduced from 5% to 1%, scrap factor in suture business has been reduced from 33% to 16%, inventory levels in the Vicryl sutures family have been reduced from 6 weeks to 2.5 weeks.

Ethicon Juarez has achieved the Integration Level in Lean (2010). This is the highest of the four levels in the J&J Lean Maturity Assessment and because of this achievement; Juarez is considered as a world class plant.

ME² is a program that focuses on maximizing the efficiency of production processes, resources and confidence in the equipment, effective use of man power, operating cost optimization and quality of the products. ME² has been a key enabler of our Lean Manufacturing strategy. As a result of this program, last year Ethicon achieved, on average, a 14% OEE improvement, Maintenance cost as percentage of CARV (Capital Asset Replacement Value), of 3.4%, and spare part inventory of 1.6% of CARV. Within J&J, an assessment system for ME² progress is utilized. Per the last assessment Ethicon Juarez reached the Performer level (2011), which is the highest of the four levels.

QUALITY & COMPLIANCE

- o Internal process deviations had a 13% reduction in 2011 versus 2010
- o Zero Recalls since 2001 in the Suture Business Unit.
- o Do it right, Do It Better (Hazlo bien, Hazlo mejor) initiative the purpose of this program is to ensure everything we do on the day-to-day is done right the first time
- o As a result of all the efforts Ethicon has achieved in terms of product quality and continuous improvement mindset, several agencies and organizations like FDA (USA), ANVISA (Brazil), BSI (ISO) and J&J have done quality management audits with ZERO Observations

PEOPLE

- o 30% Productivity Improvements have been achieved in the last 2 years with 3,175 suggestions proposed and implemented by the production Associates since 2008
- o Turnover annual rate at 7.2% versus 39.48% turnover rate for medical industry in Juarez city
- o 95% favorable response on the people engagement index since 2009 (Credo survey)

COST

- o Cost Improvement projects 5.7 million (2009-2011)

CUSTOMER SERVICE

- o 5 Days reduction in total Cycle time has been achieved during the last three years and $3.3 MM in inventory reduction was also achieved last year.
- o Over 97% Line Item Fill rate during last three years.

VALUE STREAM

- o Visual Management was also implemented throughout the manufacturing floor

- o Delivery and implementation of Kaizen events for the last years as part of the continuous improvements are of 58 Kaizen events in 2008 (implementation year) and 234 for last year (2011).

COMMUNITY SERVICE

- o Woman Leadership Initiatives (WLI), Hispanic Organization for Leadership & Achievement (HOLA) and "Comite de la Comunidad" (Community Committee) affinity groups were launched at the facility to help management achieve business impact activities. Some of the activities that are coordinated throughout the year are:
  - "Un dia en Ethicon" (One day in Ethicon), where 100 associates' kids come to the plant and experience a work day like their parents do,
  - "Banco de Leche" (Milk Bank) where 149 infants from the Community receive a daily milk portion on their first year of life,
  - Blood Chemistry where salary group has the opportunity to receive a standard blood analysis to make awareness on health problems.
  - United Way Initiatives
  - Relationship conferences
  - "Casas por Cristo" (Houses for Christ) where a house is constructed in a two-day period for a family with low resources.
  - Race for the cure employees participation
  - Juarez Competitiva is a private sector / government initiative to attract new investment to the area and help repair the city's tarnished image where associates participated in events in support of four key themes: family-oriented events, the Medical Industry and Services, International Trade, and Johnson & Johnson pavilion
  - Ethicon joined local hospitals and physicians to ensure that six bariatric procedures, including hospitalization, medication and medical services, were provided at no cost to low income residents.
  - Clean up our city initiative

AWARDS

2008 - Johnson & Johnson Worldwide sustainability Award
2008 - Johnson & Johnson Collaboration Excellence Award Worldwide Supply Chain
2011 Marketing Mastery Launch Award
2011 - Johnson & Johnson Collaboration Excellence Award Worldwide Supply Chain
2011 IndustryWeek Best Plant Winner

For more information contact:
John Schneider - Plant manager
Juarez, Mexico
(915) 791-3625
Jschnei1@its.jnj.com

REXAM

Beverage Can South America - Águas Claras

Corporate Overview

Rexam is a global consumer packaging company with a turnover of £4.9 billion in 2010. They employ some 22,000 people in more than 90 plants and offices in 20 countries around the globe. Headquartered in London, England, Rexam is an established member of the FTSE 100, a global leader in the manufacturing of beverage cans and one of the world leaders in rigid plastic packaging. They are business partners to some of the world's most famous and successful consumer brands such as Coca-Cola, InBev and PepsiCo, and blue chip global brand owners such as Proctor & Gamble and GlaxoSmithKline.

Rexam's vision to "be the best global consumer packaging company" is strongly supported by Rexam's four embedded core values: Continuous Improvement, Trust, Teamwork and Recognition. They have been committed to a path of Lean Enterprise across all operations as a way of putting values into practice for over ten years. This approach is an essential part of what they call the "Rexam Way."

Continuous Improvement Process

Rexam has operated a global system of Lean Enterprise since 2004, structured around building capability in specific tool sets of culture, VSM, 5S, TPM, SMED and six sigma. Progress is reviewed annually and awarded a merit of bronze, silver or gold depending on performance levels. Águas Claras Plant achieved gold level at their first assessment in 2004, and consecutively in 2005, 2006 and 2007.

In 2008, Rexam's best plants were invited to participate in a more sophisticated assessment known as "Beyond Gold," where the implementation of Lean systems was appraised. Again, a system of merit is used to recognize achievements starting at emerald, then sapphire, and finally diamond. Águas Claras has ranked at diamond level since the first Beyond Gold protocol assessment was made.

Rexam has made more than 166 improvement implementations based on SMED and Kaizen methods since 2004, and value stream maps are revised twice a year with the gap analysis resulting in a hopper list of main projects and actions linked to business goals.

They measure the effectiveness of actions and have increased from 73.2% in 2006 to 92% in 2010. The 5S Program, introduced in 2004, involves all employees through internal audits with a robust management control system. The commitment to continuous improvement through Lean Enterprise has led Águas Claras to be the first Rexam Can site in South America to apply for the Shingo Prize.

The Águas Claras Plant

The Águas Claras Plant

The plant, based in the Brazilian state of Rio Grande do Sul, started operations in 2002. It is one of the 12 facilities that make up the manufacturing sector known as Beverage Can South America (BCSA).

Of the 111 people on site, 95 are involved in plant operations and management and 16 are responsible for logistics and distribution. The 167,000-sq. ft. facility has one aluminum can manufacturing line that is fully dedicated to customers such as Coca-Cola, Ambev, Schincariol and Femsa, with a total production of 3.6 million cans a day. The plant operates 24 hours a day, 7 days a week, and the need for on-demand uptime drives a strong Lean culture. The plant's OEE performance improved from 76% in 2004 to 85% in 2010.

The Águas Claras plant has an established decision-making culture based on six sigma analysis of manufacturing processes. The plant has been at the forefront of Rexam's Lean Enterprise success for many years now and is seen as a benchmark for other Rexam sites in South America and across the world to follow.

Product

The site is capable of producing two sizes of beverage cans, 12oz and 16oz, for nine clients, resulting in over 101 active label options. The ability to continuously improve changeover times to meet smaller lot size requirements and make faster deliveries is at the heart of maintaining their competitive advantage.

To make a beverage can, the key process steps involve stamping and drawing aluminum discs from rolled stock to form a can shape. The cans are then washed before being labelled using in-line screen printers. This plant was designed with a quality guarantee concept at every stage of the production process, without the need for final inspection for quality. Speed and quality are the key drivers of process performance and our QAS (Quality Assurance System), RSVIEW (Rockwell Process Monitoring) system and RSBizWare (Rockwell Database System) provide real-time production data and process monitoring capability.

Achievements

Lean Enterprise Achievements:
- Gold status 2004-2007 becoming the only plant to achieve gold status since the first year of assessment
- Beyond Gold diamond status 2008-2010
- Becoming the first plant in Rexam to win this certification
- Rexam's Best Lean Business in 2008
- Rexam BCSA Best Lean Enterprise 2010, Building a Winning Organization category

Rexam PLC
4 Millbank
London SW1P 3XR
United Kingdom

REXAM

Beverage Can South America - Águas Claras

- Rexam People Development Award 2007 - for the excellence of the plant's programs in developing the skills and competencies of its team across the region
- PGQP (Local Quality and Productivity Program) achievements: Bronze level in 2004, Silver level in 2005
- ACIVI (Commercial and Industrial Association of Viamão City): Silver Quality Award, 2005
- FEBRAMEC (Brazilian Fair of Mechanical and Industrial Automation) achievements: awarded in 2008 for projects related to energy consumption and gas emissions. Repeat award in 2010 for the usage of the heat from oven exhaust chimneys in the can washing and water recycling process.

Safety & Environmental
- ISO 14001: 2004 Environmental accreditation OSHAS 18001: 2007 Safety accreditation
- Rexam Risk Management Best Practice award for achieving best practice level in Health & Safety and Environmental categories, in 2005
- HACCP and GMP accreditation from the National Food Security Program in 2009 – (Hazard Analysis and Critical Control Points, and Good Manufacturing Practices)
- Annual safety training for all employees
- Monthly safety visits from management covering all areas of operations, warehousing and administrative areas
- "Elimination of Risk Points" Program since 2006 has identified and treated more than 8,000 potential accidents and unsafe conditions
- Zero Lost Time Accidents since 2009
- Behavioral Based Safety Program began in 2010 and has achieved a 100% rate of management engagement
- Zero Environmental Accidents since 2009

Quality
- ISO 9001:2008 Quality accreditation
- Assured quality through annual customer audits
- Two black belts (plant manager and TPM technician)
- 27 senior team members are green belts
- Another eight members of the senior team are training to be green belts
- QAS (quality assurance systems) and QAS Minitab real-time SPC process monitoring covers 100% of production

Employee Morale
- Rexam is dedicated not only to people's quality of life, through a consolidated "Program of Wellbeing," but also to their career and personal development. All 111 employees at the plant are engaged in twice yearly performance appraisals and objective setting, and 100% of people have had personal development plans since 2007

The Águas Claras Plant

- Every quarter, the best employees are recognized for their engagement in all safety and Lean Programs
- Plant "Good Day" Program donates money to help social institutions, whenever safety, quality and production records are achieved
- Annual events are held for all employees to discuss targets, process improvements, suggestions, best manufacturing and administrative practices, and to set up working groups to treat matters of engagement
- They invest in and stimulate a two-way open communication between employees and the leadership team, through constant alignment meetings, a suggestion program called "Open Channel" and other events
- Also, employees can count on established corporate communication channels, such as monthly newspapers, posters, special campaigns and intranet news to be informed of company guidelines
- Absenteeism rate of 1.52% since 2007

Delivery Performance and Cost Reductions
- OTIF performance has remained consistently above 85% since 2008
- Assured supply quality through annual audits with major suppliers
- Inventory turnover fell 11% since 2009
- Electricity consumption fell by 10% since 2007
- Thermal energy consumption fell by 12% since 2007
- Spoilage has reduced by 43% since 2003
- Over Varnish Usage fell by 22% since 2007
- LPG consumption fell by 6.5% since 2007
- Washer chemical consumption has reduced 35% since 2006

For more information contact:
Jon Alder
Director, Group Lean Enterprise
Direct Line +44 (0)20 7227 4197
Mobile +44 (0)7786 197474
Email: jon.alder@rexam.com

Rexam PLC
4 Millbank
London SW1P 3XR
United Kingdom

Bibliography

Adler, P. S. The learning bureaucracy: New United Motors Manufacturing, Inc. In B. M. Staw and L. L. Cummings (eds.), *Research in Organizational Behavior*, Greenwich, CT: JAI Press, April 1992. Available at www.bcf.usc.edu/~padler/research/NUMMI(ROB)-1.pdf

Conant, D. *Saving Campbell Soup Company*. February 11, 2010. Available at http://www.gallup.com/businessjournal/125687/saving-campbell-soup-company.aspx.

Deming, W. E. *Large List of Quotes by W. Edwards Deming*. Washington, DC: The W. Edwards Deming Institute, 2017. Available at https://blog.deming.org/w-edwards-deming-quotes/large-list-of-quotes-by-w-edwards-deming/.

Deming, W. E. *Out of the Crisis*. Cambridge, MA: Massachusetts Institute of Technology, 1988.

Covey, S. R. *7 Habits of Highly Effective People*. New York, NY: Simon & Schuster, Inc., 1989.

Dillon, A. P. and Shingo, S. *A Revolution in Manufacturing: The SMED System*. Boca Raton, FL: CRC Press, 1985.

Duerr, E. C., Duerr, M. S., Ungson, G. R., and Wong, Y.-Y. Evaluating a joint venture: NUMMI at age 20. *International Journal of Business and Economy*, 2005, 6(1), 111–135.

Imai, M. *Kaizen: The Key to Japan's Competitive Success*. New York, NY: McGraw-Hill, 1986.

Imai, M. *Gemba Kaizen: A Commonsense Approach to a Continuous Improvement Strategy*. 2nd edn. New York, NY: McGraw-Hill, 2012.

Kiley, D. Goodbye, NUMMI: How a plant changed the culture of car-making. April 2, 2010. Available at www.popularmechanics.com/cars/a5514/4350856/

Langfitt, F. The end of the line for GM-Toyota joint venture. March 26, 2010. Available at www.npr.org/templates/story/story.php?storyId=125229157

Liker, J. K. *The Toyota Way*. New York, NY: McGraw-Hill, 2004.

O'Reilly III, C. A. and Pfeffer, J. *Hidden Value: How Great Companies Achieve Extraordinary Results with Ordinary People*. Boston, MA: Harvard Business School Press, 2000.

Nadler, G. and Hibino, S. *Breakthrough Thinking*. Rocklin, CA: Prima Publishing, 1998.

Plenert, G. *International Management and Production Methods: Survival Techniques for Corporate America*. Blue Ridge Summit, PA: Tab Professional and Reference Books, 1990.

Plenert, G. *Plant Operations Deskbook*. Homewood, IL: Business 1 Irwin, 1993.

Plenert, G. *World Class Manager*. Rocklin, CA: Prima Publishing, 1995.

Plenert, G. *The eManager: Value Chain Management in an eCommerce World*. Dublin, Ireland: Blackhall Publishing, 2001.

Plenert, G. *International Operations Management*. Copenhagen, Denmark: Copenhagen Business School Press (republished in India by Ane Books), 2002.

Plenert, G. *Reinventing Lean: Introducing Lean Management into the Supply Chain*. Amsterdam, the Netherlands: Elsevier Science, 2007.

Plenert, G. *Lean Management Principles for Information Technology*. Boca Raton, FL: Taylor & Francis Group, CRC Press, 2012.

Plenert, G. *Strategic Continuous Process Improvement: Which Quality Tools to Use, and When to Use Them*. New York, NY: McGraw-Hill, 2012.

Plenert, G. *Supply Chain Optimization through Segmentation and Analytics*. Boca Raton, FL: Taylor & Francis Group, CRC Press, 2014.

Plenert, G and Cluley, T. *Driving Strategy to Execution Using Lean Six Sigma: A Framework for Creating High Performance Organizations*. Boca Raton, FL: CRC Press, 2013.

Plenert, G. and Hibino, S. *Making Innovation Happen: Concept Management through Integration*. Delray Beach, FL: St. Lucie Press, 1997.

Plenert, G., Kirchmier, B., and Quinn, G. *Finite Capacity Scheduling: Optimizing a Constrained Supply Chain*. Self-Published, 2014.

Senge, P. M. *The Fifth Discipline: The Art & Practice of the Learning Organization*. 2nd edn. New York, NY: Random House, Inc., 2006.

Shingo Institute. *Shingo Model Handbook*. Downloadable at www.shingo.org/model

Shingo, S. *Zero Quality Control: Source Inspection and the Poka-Yoke System*. Boca Raton, FL: CRC Press, 1986.

Shingo, S. *Non-Stock Production: The Shingo System of Continuous Improvement*. Portland, OR: Productivity Press, 1988.

Shingo, S. *The Shingo Production Management System: Improving Process Functions*. Portland, OR: Productivity Press, 1992.

Shingo, S. *Kaizen and the Art of Creative Thinking: The Scientific Thinking Mechanism*. Bellingham, WA: Enna Products Corporation and PCS Inc., 2007.

Shingo, S. *Fundamental Principles of Lean Manufacturing*. Bellingham, WA: Enna Products Corporation, 2009.

Shingo, S. and Dillon, A. P. *A Study of the Toyota Production System: From an Industrial Engineering Viewpoint*. Boca Raton, FL: CRC Press, 1989.

Taylor, A. How Toyota lost its way. *Fortune Global 500*. July 12, 2010. Available at http://archive.fortune.com/2010/07/12/news/international/toyota_recall_crisis_full_version.fortune/index.htm

Womack, J. P., Jones, D. T., and Roos, D. *The Machine That Changed the World: The Story of Lean Production—Toyota's Secret Weapon in the Global Car Wars That Is Now Revolutionizing World Industry*. New York, NY: Simon & Schuster, Inc., 1990.

Young, J. *Steve Jobs: The Journey is the Reward*, Lynx Books, 1988.

Index

About the Editor

Dr. Gerhard Plenert, former director of executive education at the Shingo Institute, has more than 25 years of professional experience in organizational transformations, helping companies and government agencies strive for enterprise excellence by utilizing the *Shingo Model* to drive cultural transformations. Dr. Plenert is an internationally recognized expert in supply chain management; Lean/Six Sigma; IT, quality and productivity tools; and working with leading-edge planning and scheduling methods. He has literally "written the book" on leading-edge supply chain management concepts, such as finite capacity scheduling, advanced planning and scheduling, and world-class management.

His experience includes significant initiatives with Genentech, Johnson & Johnson, Aerojet Rocketdyne, Shell, Aramco, Sony, Cisco, Microsoft, Seagate, NCR Corporation, Ritz-Carlton, the U.S. Air Force, and numerous branches of the U.S. Department of Defense. In addition, Dr. Plenert has consulted for major manufacturing and distribution companies such as Hewlett-Packard, Black & Decker, Raytheon, Motorola, Applied Magnetics, Toyota, AT&T, IBM, and Kraft Foods. He has also been considered a corporate "guru" on supply chain management for Wipro, AMS, and Infosys, and a Lean/Six Sigma "guru" for the U.S. Air Force and various consulting companies.

With 14 years of academic experience, Dr. Plenert has published over 150 articles and 18 books on Lean, supply chain strategy, operations management, and planning. He has also written MBA textbooks and operations planning books for the United Nations. Dr. Plenert's ideas and publications have been endorsed by people like Stephen Covey and companies such as Motorola, AT&T, Black & Decker, and FedEx. His publications are viewable at www.gerhardplenert.com.

Dr. Plenert previously served as a tenured full professor at California State University, Chico; a professor at BYU, BYU–Hawaii, University of Malaysia, and the University of San Diego; and has been a visiting professor at numerous universities all over the world. He earned degrees in math, physics, and German; and he holds an MBA, MA in international studies, and PhD in resource economics (oil and gas) and operations management degrees. Dr. Plenert continues to serve as a Shingo examiner.